One Physicist's Guide to Nuclear Weapons

A global perspective

One Physicist's Guide to Nuclear Weapons

A global perspective

Jeremy Bernstein
Professor Emeritus, Stevens Institute of Technology

IOP Publishing, Bristol, UK

ISBN 978-0-7503-1308-7 (ebook)
ISBN 978-0-7503-1309-4 (print)
ISBN 978-0-7503-1310-0 (mobi)

DOI 10.1088/978-0-7503-1308-7

Version: 20160701

IOP Expanding Physics
ISSN 2053-2563 (online)
ISSN 2054-7315 (print)

British Library Cataloguing-in-Publication Data: A catalogue record for this book is available from the British Library.

Published by IOP Publishing, wholly owned by The Institute of Physics, London

IOP Publishing, Temple Circus, Temple Way, Bristol, BS1 6HG, UK

US Office: IOP Publishing, Inc., 190 North Independence Mall West, Suite 601, Philadelphia, PA 19106, USA

Contents

Preface

Until the collapse of the Soviet Union the issue of nuclear weapons seemed in some sense done and dusted. As Stanley Kubrick once said to me when he was thinking about *Dr Strangelove*[1], there seemed to be less interest in nuclear weapons than in city government. There was a doctrine of mutually assured destruction which one had a vague feeling was keeping the peace. Gone were the days when in school we had drills in which we were taught to hide under our desks in case of a nuclear attack. There was no above-ground testing. The last such test was conducted by the Chinese in 1980. We had simply repressed the thought of what an actual nuclear explosion looked like. This was good in that the issues of fallout, with the attendant cancer rates, were no longer relevant. But it was bad in the sense that there was less and less a sense of what these weapons were about. As Kubrick said, they were like your own mortality. You knew it was inevitable, but there was a very effective method of denial. Since we did not see them, they had effectively ceased to exist.

The notion of 'third world' countries like Iran and North Korea, and even India and Pakistan, having nuclear weapons seemed absurd. They could not possibly have the technology or the resources. The existential urge for a country like Pakistan to have nuclear weapons was not taken fully into account. Now we are in a very different world. We find ourselves having to deal with third world countries with growing nuclear arsenals. I have been thinking about this for a long time. I saw the first boosted nuclear weapon tested in the desert of Nevada in 1957. It had over twice the explosive yield of the bomb that flattened Nagasaki. I was given the plutonium pit of a nuclear weapon to hold. The effect on me was to give me a sense of power. It took a bit of time before I realized how absurd this was. When I began to write professionally I saw that there was something I could do, since as a physicist I had some understanding of the technology. I could explain. I could illuminate the history. I could create the panorama. And I could try to do this in a manageable form.

I studied in detail the program in Iran. I realized at once that Iran is a country with a very significant scientific culture. The Shah donated millions to MIT and sent students there to be educated. I studied the curriculum and faculties at some of the important technological institutions. I saw that they were publishing in international journals and holding conferences in Teheran. They were producing their own scientists. When it comes to scientists they do not need our help. I studied the nuclear weapons development in Pakistan, which was, on the theoretical side, led by a friend of mine. I saw that a very small number of talented people with the right motivation and support could create nuclear weapons. This led me to wonder why the Germans failed. They had some of the greatest physicists in the world and the

[1] S Kubrick 1964 *Dr Strangelove or: How I Learned to Stop Worrying and Love the Bomb* (film).

support of the government, yet they failed. Why did Pakistan with its limited resources succeed? Why did North Korea with even more limited resources succeed? How close have the Iranians come? These are questions we would not have asked—I would not have asked—a few years ago. That has been my concern. I am afraid that it has become our concern.

Foreword

'This is the office of Jeremy Bernstein, who owes his not inconsiderable reputation to his judicious choice of collaborators'. The note which the author of this collection of essays put on his door in the 1960s displays some of his qualifications as a guide to nuclear weapons and their deployment. Jeremey Bernstein, himself an excellent physicist like his collaborators, knew most of the scientists who played key roles in the development of nuclear weapons at Los Alamos. He also writes entertainingly, as those who have read his pieces in the *New Yorker* and elsewhere know, without losing sight of the gravity of his subject matter.

Bernstein was not involved in the weapons program, but he is one of the dwindling number of people who have witnessed the terrifying force of a nuclear explosion. This leads him to reflect that 'it is a pity that people who suggest [research to improve nuclear weapons] have never seen one explode'.

A major focus of these essays is on the early development of nuclear weapons and how they work, which Bernstein describes in a way that is accessible to lay readers, with many personal touches and insights. Moving from the Americans and their allies to the scientists who worked in the German nuclear program during World War II, he asks: were they trying to make a weapon? If so, how close did they come? If they had succeeded, would they have turned it over to Hitler? This book offers a great deal to interest those who have studied these perennially fascinating and still controversial questions, as well as those who have not.

Bernstein's description of how Pakistan, Iran, North Korea, and Israel acquired nuclear weapons is deeply disturbing. The worrying possibility of ISIS buying a bomb is only partly mitigated by his belief that they would not be able to obtain an effective means of delivery.

Other essays touch on the early Soviet nuclear weapons programme, and the extent to which it benefitted from information provided by Klaus Fuchs about work at Los Alamos. Bernstein believes that President Truman's conviction that the Soviets could never make a bomb led him to rush into—mistakenly—authorizing the development of the hydrogen bomb when he learned that they had done so. An essay on J Robert Oppenheimer, based on personal knowledge and careful research, not only casts light on his role in the debate over whether to develop the hydrogen bomb, which cost him his scrutiny clearance, but also on his personality (why did he claim that J, for Julius, stood for nothing?) and the question of whether he was ever a member of the Communist Party.

Professor Sir Chris Llewellyn Smith, FRS
Director of Energy Research, Oxford University
President, SESAME Council
Director General of CERN 1994–98

Acknowledgements

Portions of this book were adapted from three short works by the author: 'At Los Alamos', 'How Iran got the bomb', and 'Oppenheimer's lives', published electronically by *Now and Then Reader* at nowandthenreader.com. In addition, portions of this book were adapted from articles by the author that appeared in the *New York Review of Books*, on Los Alamos, Iran, India, Pakistan, and North Korea, specifically: 'Pakistan and India' (2013) and 'A mushroom cloud recalled' (2010).

I have been aided by a number of colleagues whom I would like to acknowledge here: David Albright, Lowell Brown, Tom Cochran, Norman Dombey, Freeman Dyson, Richard Garwin, Roy Glauber, Gerald Holton, Cameron Reed and Chris Llewellyn Smith. Above all I am grateful to Ken Ford for his editorial help.

I would like to thank Ken Ford for a very careful reading and editing of the manuscript, and Jessica Fricchione and Dan Heatley for their help in preparing the text. Many thanks also to Sir Chris Llewellyn Smith for his detailed comments and suggestions, and to Robert Gallucci for reading the manuscript.

Author biography

Jeremy Bernstein

Jeremy Bernstein is a professor emeritus of physics at the Stevens Institute of Technology in Hoboken, New Jersey. He was a staff writer for the *New Yorker* from 1961 to 1995. He was an adjunct professor at the Rockefeller University and a former vice president of the Board of Trustees of the Aspen Center for Physics of which he is now an Honorary Trustee. He is also a fellow of the New York Institute for the Humanities. He has held appointments at the Institute for Advanced Study, Brookhaven National Laboratory, CERN, Oxford University, the University of Islamabad, and the Ecole Polytechnique. He has written some fifty technical papers and several technical monographs; including *Kinetic Theory in the Expanding Universe*, *An Introduction to Cosmology*, *Secrets of the Old One: Einstein*, 1905 and *Quantum Leaps*. In addition, he has written fourteen books on popular science and mountain travel. They include *Albert Einstein*, which was nominated for a National Book Award, *Three Degrees Above Zero*, *In the Himalayas*, *Mountain Passages*, *Science Observed*, *The Tenth Dimension*, *Cosmological Constants*, *Quantum Profiles*, *Cranks, Quarks, and the Cosmos*, *Hitler's Uranium Club-a history of the German attempts to build an atomic bomb*, *A Theory for Everything*, *Dawning of the Raj* and *The Merely Personal: Observations on Science and Scientists*, and a biography of Robert Oppenheimer entitled, *Oppenheimer: Portrait of an Enigma*, two books on nuclear weapons, *Plutonium* and *Nuclear Weapons: What you Need to Know*. He has won the AAS-Westinghouse and the US Steel-American Institute of Physics science writing prizes; the latter twice. He has also won the John Case Award of the American Alpine Club for his writing about the mountains, the Brandeis Creative Arts Medal for general non-fiction and the Britannica Award for the dissemination of knowledge. The latter included a lecture on cosmology at the Tribhuvan University in Kathmandu. He has also won the Gemant Award of the American Institute of Physics for contributions that link physics to the arts and humanities, and the Ralph Coats Roe medal of the American Society of Mechanical Engineers for contributions to the public understanding of science and technology, and the Hartley Burr Alexander Award of the University of Nebraska. Most recently he has won the Burton Award of the American Physical Society for his work on science communication. He has taught non-fiction writing as a Ferris Professor of Journalism at Princeton University, most recently in the Fall of 1992. He has had articles published in most of the national magazines. He was a National Sigma Xi lecturer, and a member of the Board of Editors of the *American Scholar* for which he wrote a column titled 'Out of My Mind'. He was a fellow of the American Physical Society, was a Benjamin Franklin Fellow of the Royal Society of the Arts and a member of the French and American Alpine Clubs. He was born in Rochester, New York on 31 December 1929, and educated at Harvard University, from which he received his BA, MA and PhD degrees.

Chapter 1

My up-close experience

This is the way the world ends
This is the way the world ends
This is the way the world ends
Not with a bang but a whimper.
T S Eliot[1]

I graduated from Harvard having majored in mathematics in 1951. I got my PhD in physics in 1955. I needed a job and Gerry Holton, a physics professor who had become a friend, made a suggestion. On the Harvard campus there was a cyclotron. It had a relatively modest energy. Its operation was simple enough, so it could be used by graduate students as well as faculty members. There was a position there called 'house theorist'. The occupant was moving on to a university, so it was open. Holton recommended me and I got the job. I inquired as to what it consisted of and was told that my only formal duties would be to try to answer questions put by the experimenters. Otherwise I could do my own work. Since the experimenters were people like Norman Ramsey and Ed Purcell, eventually Nobel Prize winners, I thought that if anything the shoe would be on the other foot and I would have plenty of time to do my own work. My main interest was in the interpretation of the experiments that were being done at Stanford by Robert Hofstadter. He was scattering electrons from protons, deuterons and other nuclei. I had written my thesis dissertation on the deuteron, which explains my interest.

After two years my appointment was over and I had to look for a new job. Without much hope of success I applied to the Institute for Advanced Study and was incredibly happy when I was accepted for the fall of 1957. This left the summer. What then happened was quite unexpected. Let me give a little background.

[1] T S Eliot 1925 The hollow men *Poems* 1909–1925.

This was the height of the Cold War and the weapons laboratories such as Livermore and Los Alamos were actively recruiting. Indeed, in the spring of 1954 while I was still working on my thesis I had an unsuccessful interview in Washington with Edward Teller, who was recruiting for Livermore. He made an odd comment about preferring physics to politics. It was only much later that I realized that my interview had taken place the day after he had testified against Robert Oppenheimer at a hearing that resulted in Oppenheimer's loss of his clearance. What I also did not know in 1957 was the very active role that many of the professors I knew had played in creating the bomb.

To make a partial list: Norman Ramsey had been involved in selecting the airplane used to deliver the bomb over Japan. He also signed off on the plutonium device that destroyed Nagasaki. Roy Glauber, who when I first knew him at Harvard was a young assistant professor, had been the youngest member of the technical staff at Los Alamos. He had been recruited even before he had received his bachelor's degree from Harvard. His college roommate Ted Hall was the second youngest staff member. He was also one of the three known Soviet spies at Los Alamos. At MIT I had gotten to know Viki Weisskopf, who was Hans Bethe's second in command in Los Alamos's theory division. Of course the chemist James Conant who was president of Harvard had been one of the people who ran the whole enterprise. Of more relevance to me was Ken Bainbridge. He had selected the Alamogordo test site and was in charge of arming the device that was successfully tested there on 16 July 1945. Now he was chairman of the physics department at Harvard.

What was odd was that I never heard any of these people discussing their wartime activities. Maybe it was too close to the events or maybe there was still too much secrecy. So not knowing any of this I was surprised when Bainbridge called me into his office and asked if I would like a summer job at Los Alamos. I told him that I would and he said that he would recommend me. I was very curious about Los Alamos, which for me had an almost mystic quality—a magic mountain in New Mexico. Also, some extra money would come in handy when I got to Princeton. Although I do not remember the specifics, I must have met with a Los Alamos recruiter, who told me that I could have the job if I was able to obtain the relevant clearance.

In 1947 the new Atomic Energy Commission introduced a Personnel Security Questionnaire (PSQ) to decide levels of clearance. The levels were appropriately labeled P, S, and Q, Q being the highest. A person with Q clearance was entitled to information about nuclear weapons on a 'need to know' basis. In due course I had to supply the FBI with a list of all my residences going back some ten years. Much later, using the Freedom of Information Act, I managed to get my file, but it was so heavily redacted that I did not learn anything. I was rather worried about a great aunt. She was a subscriber to the *Daily Worker* and spoke in dark tones about the 'bosses'—I presume she meant the capitalist bosses. Either the FBI overlooked her or they decided that she was harmless, because I got my clearance.

While I remember a great deal about that summer, what I do not remember is how I got to Los Alamos. It was surely not by car. I had left my somewhat aged

Morris Minor convertible at my parent's house in upstate New York. Possibly I got there by plane, which would have meant landing in Albuquerque and then taking some transportation to Santa Fe. Perhaps I took the train, which would have let me off in Lamy, where I might have found my way to 109 East Palace Avenue in Santa Fe, the address of an office from which the wartime scientists were taken up the hill to the laboratory. I simply cannot remember. I do remember arriving at the guard station at Los Alamos and being fitted out with credentials. You needed a pass with your picture on it to enter any of the technical areas. The Los Alamos of 1957 was a much more guarded place than it had been during the war. Indeed, Oppenheimer had assigned Robert Serber and his wife to go into Santa Fe on a Saturday night and plant rumors at bars that what was happening on the hill had to do with submarines. No one showed the slightest interest. By 1957 everyone knew what was happening on the hill. Klaus Fuchs had been convicted of espionage in 1950. The security at Los Alamos was very tight.

As a single person I was assigned quarters in a dormitory that must have been left over from the war. It was pretty basic. I immediately acquired a bicycle, which became my principal means of transportation for the rest of the summer. The next day I rode my bicycle to the building where T Division (T for theory) was located. I showed my pass to a series of guards and found my way to my office. I discovered that I was sharing it with Ken Johnson, whom I had known since my graduate school days. He had written a first rate thesis and had been kept on as a post doc in the Harvard department. He was scheduled to go to the Niels Bohr Institute in Copenhagen in the fall. It soon became clear to us that no one had any work for us, so we were free to do anything we wanted. I had come with a problem of how to determine the parity of the pi zero using aspects of its two-photon decay. (The same question has now arisen with the parity of the Higgs boson.) I was sort of stuck on the mathematics, so we decided to do it together. Ken was a very powerful mathematician and he proved results of very great generality. We wrote a paper and submitted it to the director of the theoretical division, Carson Mark (figure 1.1), to see if we could publish it with a Los Alamos imprimatur. Mark was very pleased, because he did not want Los Alamos to only have the reputation as a bomb factory.

Of course in 1957 it *was* a bomb factory. It was the height of the Cold War and Los Alamos—and Livermore—were churning out designs for devices small enough

Figure 1.1. Carson Mark. Courtesy of Los Alamos National Laboratory

to fit into intercontinental missiles. These were being tested above ground in Nevada in a series that was called Operation Plumbbob. I knew in a general way about the tests, but nothing about the specifics. I had no idea that I would have an involvement at the end of the summer. We had afternoon teas—something that Oppenheimer had introduced during the war—and I am sure that most of the people there were working on weapons. Since I had no 'need to know', no one told me anything. There is one memory of those teas which is ineluctable. James Tuck was a British physicist who had come with the British delegation to Los Alamos during the war. After a few career changes he had decided to come back to Los Alamos to work on controlled nuclear fusion, which was in 1957 a classified subject. I recall him as a somewhat Falstaffian figure who wore shorts. The first time I saw him at tea he said, 'The days of the great Los Alamos teas are over.' Once he could have tea with Niels Bohr or Fermi; now he had to make do with the likes of us.

The controlled nuclear fusion program had attracted a very high level of consultant. Among them was Francis Low (figure 1.2).

I had met Francis briefly during the year that he had been a visiting professor at MIT. After the summer he was going there permanently. He was a hero of mine. His work with Geoffrey Chew on meson theory was vastly superior to what I had used in my thesis. He and Murray Gell-Mann had made a study of quantum electrodynamics at short distances that introduced techniques that are still basic to modern quantum field theory. At Los Alamos I learned that Francis was a devoted tennis player and we began to play together. We even played doubles on the Los Alamos team in a match against a much superior team from Santa Fe. I also learned that in the war he had been a member of the 10th Mountain Division, which had fought in Italy. He told me that he could have obtained a deferment to work in a place like Oak Ridge, but he felt that his knowledge of physics was so meager that he would not have been of much use. This did not exactly explain why he had volunteered for the mountain troops. I got a sense of his derring-do when he asked me to transport him to the lab on the handlebars of my bicycle. He got on, noting that he always did

Figure 1.2. Francis Low. Photograph by Kurt Gottfried, courtesy AIP Emilio Segre Visual Archives, Gottfried Collection.

this when he was a kid. I remarked that he must have been somewhat smaller then. We followed a regular tennis regimen until the middle of August, when Francis announced that he would be away watching bomb tests in Nevada. I was very surprised and asked if he had been working on weapons. He said no, but that Carson Mark had invited him to observe a test. I asked if there was any chance that I could go too and Francis said that I would have to ask Carson.

When I went to see Carson he said that I could come with them, provided that I paid my way. What this meant in practice was that I would pay my round-trip airfare from Albuquerque to Las Vegas. I said that I would be very willing to do this. Thus it came to pass that on the morning of 30 August the three of us took off from the small airstrip at Los Alamos on a commuter flight that took us to Albuquerque.

Nothing had been said, but I realized that I was about to enter into the secret world—the need-to-know world. I decided that under no circumstances would I ask any questions. I had no legitimate need to know. For example, I had no idea of our itinerary. I knew that we would have to get from Las Vegas to Mercury, Nevada, the location of the test site, some sixty five miles northwest of Las Vegas. The fact that nuclear weapons were being exploded above ground, which dumped thousands of kilocuries of radiation into the atmosphere that close to a major American city, is an illustration of the craziness of that time. I also knew that blackjack had become part of the Los Alamos gestalt. In 1956 four American soldiers stationed at the Aberdeen Proving Ground near Baltimore had published a paper in *Journal of the American Statistical Association* entitled 'Optimum strategy in blackjack'. This and their subsequent monograph explained how to play blackjack using casino rules in such a way as to optimize your chances. This came to the attention of theorists at Los Alamos, who programmed the MANIAC computer to run tens of thousands of hands to see if the strategy actually worked. Once they were satisfied that it did, a little card was given to Los Alamos people who went to Mercury showing how to play the game. Francis had made a study of the method and concluded that if you were lucky you might make an amount of money that was about equal to the federal minimum wage. Hence I was sure that we would visit a casino, but I was surprised that this was our first stop. We had been met at the plane by a small delegation of Los Alamos people in a government car which deposited us at a casino.

The casino must have received a lot of business from people at the test site because there was a light that was turned on if the test that was scheduled for the next morning was on. The light came on and we drove to Mercury to get a few hours of sleep. The tests were scheduled for 5.30 am so as to optimize the light. Carson woke us up and we walked over to a place where a meteorologist was checking the winds. That morning a device named 'Smoky' (figure 1.3) was to be tested. Carson explained that this was a Livermore device. You could always tell because they named theirs after mountains. The Los Alamos devices were named after scientists. 'Galileo' was in a tower (figure 1.4) being readied for a test in two days. The winds were pronounced all right, so we went to a concrete bunker to await the explosion. I was surprised to find an old graduate student friend of mine named Al Peaslee there. Peaslee had taken his degree and then sort of disappeared. Here he was on the test site. He had seen several tests and his present job was to escort a British

Figure 1.3. Smoky—31 August 1957. Courtesy of the National Nuclear Security Administration. Credit: Wikimedia Commons.

Figure 1.4. The 700 foot tower for Smoky, with the lift. Courtesy of the National Nuclear Security Administration.

delegation. 'We don't tell them anything,' Peaslee explained. He instructed me as to what to do. I was to face away from the explosion and count to ten. I was also given some very dark glass to put over my own glasses. Even the reflection from the bunker walls could damage your eyes. I do not know how far from the explosion we were, but we were close enough so that I could clearly see the 700 foot tower on top of which was the bomb. I also noticed a hill in back of the tower on which there was a grove of Joshua trees. These are ubiquitous in the desert southwest and look as if they are praying. A loudspeaker gave the times until the explosion and then counted down from ten. I had turned my back and covered my eyes with the dark glass, but the bright flash still made me close my eyes. I counted to ten and then turned around. I do not really have the words to describe adequately what I saw.

The entire horizon in front of me was in turmoil. In the center was a livid red-orange cloud—a boiling cauldron. The hugeness of it was what impressed me. I had had no idea of the sheer scale of a nuclear explosion. Peaslee had prepared me for the next step. I felt a sharp and slightly painful click in my ears. This was the supersonic shock wave. At Hiroshima this produced a wind that was stronger than any known typhoon. It knocked over the kerosene cookers that the Japanese were using to make their breakfasts. This is what caused most of the fires at Hiroshima. Then came the sound—a rolling thunder. By this time the cloud had turned a sort of purple and black and hung in the air like a radioactive cobra about to strike. There was some talk of taking cover but the cloud did not come in our direction. I stood there mute. We then returned to the dormitory to get a little more sleep.

Sometime about mid-morning I heard the sound of helicopters flying. I said to Carson, 'They are flying.' He said, 'They are flying and flying,' offering no explanation. Carson had at his disposal a government car. He was going to drive to the 500 foot tower from which the next test—a Los Alamos test, Galileo—was going to be exploded. We came along. On the way there were spots in the desert in which the sand had been fused into glass. There were other places where signs were posted warning of high radioactivity, all pointing to previous tests. We got to the base of the tower. To get to its top one rode most of the way in an open lift. From there on it was a rather rickety steel ladder. It felt very insecure and the desert looked a great distance below. I had a feeling of panic, but then it occurred to me that at the top of the ladder there was a nuclear device with a yield comparable to the bomb that flattened Hiroshima—and I was worried about climbing a ladder?

The top was a flat space with about enough room for Galileo and its attendants. It was a big device with various wires coming out of it. It looked more like a diving bell than a bomb. There was a clicking noise. I had been around experimenters enough at the cyclotron to know that this was a vacuum pump. I had no idea why it was there and did not ask. Carson spoke to the crew and we went back down the elevator. I thought that the tour might be over, but Carson drove to a concrete blockhouse at the very far edge of the site. He offered no explanation.

He walked in without knocking or ringing and we followed. When I looked around I stepped back towards the door. Neatly arranged on shelves were the plutonium pits—centers of bombs—of a considerable number of atomic bombs, probably enough to destroy many cities. I had read enough about Hiroshima and

Nagasaki to know what I was looking at. When he saw me retreat Francis said quietly that if one of them went off being a few feet further away was not going to make much difference. Carson had seen my retreat too. He simply picked up one the pits and handed it so me, saying something about not dropping it. The pit was warm to the touch because of the radioactivity—and about the size and weight of a bowling ball. I do not remember its color. I did not know enough to ask the obvious question: why was it so light? A solid sphere like this of plutonium would have weighed a couple of hundred pounds. I do not know if this struck Francis too. I am sure that if I had asked this question not only would I not have received an answer, but even asking it would have been viewed unfavorably. It was then that I noticed her.

There was a large workbench at the other end of the building. Seated at it was a man with a file. He was filing something that looked to me like a white putty. I had read enough to know what this was. It was a high explosive that was going to be attached to the pit to create the implosion of the plutonium sphere. He seemed very artisanal, but next to him was a woman knitting. It was many years ago, but I think that she was making a green sweater. I do not mean to be pretentious, but I at once thought of Eliot's poem with the line 'This is the way the world ends'[2]. What was she doing there? I did not dare ask. Some years later I wanted to write about this. I sent what I had written to Francis. He said that there had been no woman. I had made it up. I then wrote to Carson. All he said was that there had been more than one married couple on the site. I was somewhat distressed. A week later, Francis called. All he said was, 'She was there.'

The next morning Galileo was tested. I knew what to expect, but was still overwhelmed. Then we returned to Los Alamos. None of us talked about the tests. For Carson it was his work and Francis and I simply did not want to talk about it. We did not talk about it when I drove with Francis, who was on his way to Cambridge, as far as Lake Forest, which is close to Chicago. I now regret that I did not talk to Francis then. He might have brought me to my senses. Somehow I felt that the experience had given me power—had made me part of the secret world. If you like, I had learned to love the bomb. During my visit to Lake Forest I was introduced to Adlai Stevenson. I somehow conveyed my feelings about the bomb to him, but before I could say much he looked at me with contempt and walked away. He had made a moratorium on nuclear tests part of his unsuccessful 1956 presidential campaign, only to be repudiated by Eisenhower, who was actually considering the possibility privately. I had said the wrong thing to the wrong person.

Over the next years I came to realize how foolish I had been, especially as I learned more and more about these tests, including the ones I had witnessed. The Plumbbob series to which Smoky and Galileo belonged was the biggest and longest series of tests ever performed in the continental United States. There were 29 tests, beginning 28 May 1957 and ending 28 September 1957. The highest explosive yield was in 'Hood', the test that took place on 5 July. The yield was 74 kilotons TNT equivalent. Keep in mind that the Nagasaki bomb's yield was about 20 kilotons. Smoky had

[2] T S Eliot 1925 The hollow men *Poems* 1909–1925.

the second highest yield, with 44 kilotons. The series, during which the total yield was about 306 kilotons —something like a tenth of the yield of one hydrogen bomb—released about 58,300 kilocuries of radio iodine into the atmosphere. This fallout was distributed all over the United States and was estimated to have caused something like 32,000 cases of thyroid cancer. Twelve-hundred pigs were exposed to the effects of the explosions. Some 18,000 servicemen also participated (figure 1.5). About 1,200 were involved with Smoky.

They watched the explosion from a distance of about 13 km. Then a unit was flown to ground zero about 15 min later. They declared that it was safe to occupy, so the rest were flown in at 5.50 am—20 min after the explosion. The exercise was completed at 9.45, which is when I heard the helicopters. A few of these men contracted leukemia later in life.

The plutonium pit that I was given to hold was so light was because it was hollow. The weapons that were being tested that summer were what are known as 'boosted'. Deuterium and tritium gas was injected into the cavity just before the explosion. I imagine that the vacuum pump I heard when we visited 'Galileo' was connected to this. When the pit is imploded and the density is increased enough to reach a super-critical mass the fission chain reaction begins. When about one per cent of the plutonium has been fissioned the temperature is raised enough so that the fusion reactions of the deuterium and tritium take place. These produce a burst of very

Figure 1.5. Soldiers waiting to march to ground zero. Courtesy of National Archives, photo no.111-SC-389297. Credit: Wikimedia Commons.

high-energy neutrons which boost the subsequent fission efficiency. That is what accounted for the large yields in some of the bombs tested that summer. There is no end to the ingenuity that has been applied to weapons design.

The last above-ground test by the United States was in 1962 and the last above-ground test anywhere was conducted by China in 1980. This is certainly a good thing. But I have only one misgiving. No one has seen a nuclear explosion for over 30 years and the number of people who have ever seen one is rapidly decreasing. Of the people I was with in Nevada in 1957, Al Peaslee died in 1976, Carson died in 1997 and Francis died in 2007. For most people nuclear weapons are an abstraction. Perhaps there should be one more explosion in the desert of Nevada to remind us of what they actually are.

Chapter 2

The United States and Great Britain in World War II

The first nuclear weapon to be tested was known as the 'gadget' (a word chosen for security reasons so that the scientists would not accidentally say 'bomb' in public). The gadget was exploded at 5.29 and 45 seconds on the morning of 16 July 1945 in a scrub desert which was part of the Jornada del Muerto—a wagon trail across New Mexico where many travelers had died. Robert Oppenheimer had called the test 'Trinity'. When, many years later, General Leslie Groves, who headed the Manhattan Project of which Los Alamos was part, asked him why the name, he said that he thought that he had been reading Donne's devotional poem, 'Holy sonnet XIV', which begins 'Batter my heart, three person'd God'[1]. Then he added, 'Beyond this, I have no clues whatever.' Kenneth Bainbridge, who was my department chairman during part of the time I was at Harvard, went over to Oppenheimer after the explosion and said, 'Now we are all sons of bitches.' Frank Oppenheimer, who was Robert's brother, and who was standing next to him when the bomb went off, told me that Robert simply said 'It worked.' The Sanskrit mantra which Oppenheimer said he recited came later. I I Rabi gave his impressions: 'Suddenly there was an enormous flash of light, the brightest light that I have ever seen or what I think anyone else has ever seen.' He is overlooking the residents of Hiroshima and Nagasaki. 'It blasted; it pounced; it bored its way into you. It was a vision which was seen with more than the eye. You would wish it would stop; altogether it lasted about two seconds. Finally it was over, diminishing, and we looked toward the place where the bomb had been; there was an enormous ball of fire which grew and grew and it rolled as it grew; it went up into the air, in yellow flashes and into scarlet and green. It looked menacing. It seemed to come toward one […] A new thing had just

[1] J Donne 1633 Holy sonnet XIV *Love Songs and Sonnets.*

doi:10.1088/978-0-7503-1308-7CH2

been born; a new control, a new understanding of man, which man had acquired over nature.'

On 6 August 'Little Boy' was dropped on Hiroshima. It was a uranium fission bomb of a design that had not been tested. Hiroshima was the test. On 9 August 'Fat Man' was dropped on Nagasaki. It was a plutonium fission bomb, essentially a clone of the gadget. These are the only two nuclear weapons to have been used, so far, in combat. But testing of new and improved designs of nuclear weapons began very soon after the end of the war. These tests were above ground. The fission bomb tests were done largely at a site at Mercury, Nevada, and I witnessed two of them. Fusion bombs were tested in the South Pacific, along with a few fission bombs. We tested above ground until 1962, as did the Soviets. The British stopped in 1958 and the French in 1974. The Chinese were the last to stop in 1980. Since then several countries have tested underground. While the fact that above-ground testing stopped more than a quarter of a century ago worldwide and in America four decades ago is certainly a good thing, there is I think a down side. The number of people who have actually seen a nuclear explosion is dwindling away. Two of my Harvard professors who became Nobel Prize winners were at Trinity. Norman Ramsey, who won his Nobel in 1989 and was born in 1915, helped assemble Fat Man on Tinian Island. I taught his class when he went to defend Harvard against Joseph McCarthy. He did so well that McCarthy offered him a job. Roy Glauber, who was born in 1925, won his Nobel in 2005. He was the youngest member of the technical staff at Los Alamos. The second youngest, by about a month, was Theodore Hall, who had been Glauber's roommate in their senior year at Harvard. It later turned out that Hall was a Soviet spy. But each year there are fewer and fewer witnesses to an atomic explosion.

It has been over half a century since the 'Bravo' test of a hydrogen bomb was performed on the atoll of Bikini—1 March 1954, to be exact. The designers, some of whom I know, claimed that a certain nuclear reaction involving an isotope of lithium would not be important. They were wrong. The bomb was twice as powerful as they had predicted. It was about a thousand times more powerful than the Hiroshima bomb. It produced a fireball that, within seconds, was three miles wide. It was visible for almost a minute on the island of Rongerik, 135 miles away. It pulverized coral, producing a grey radioactive ash. The Japanese fishing boat unluckily named the *Lucky Dragon* was fishing for tuna. It was about 80 miles away from Bikini in what was thought to be a safe area. White ash began falling on the ship. The crew of 23 had no idea what it was, but when the ship returned to port two weeks later, they were all suffering from radiation sickness and the radio operator later died. This incident caused a worldwide protest against these atmospheric tests, but we and the Soviets continued for another eight years, partly because people like Edward Teller put forward one specious argument after another about the risks of the Soviets cheating. If there was any residual doubt about our ability to detect underground nuclear tests it should have been put to rest by the 9 October 2006 North Korean test. Within days it had been determined that it was a plutonium fission device with the explosive power of a small fraction of the Hiroshima bomb, even though it was hidden underground. Later I will explain how something like this can be determined.

Lord Cherwell, when the war began,
Was plain professor Lindemann.
But now, midst ministerial cheers,
He takes his place among the peers.
The House of Christ with one accord
Now greets its newly risen Lord.
Hugh Trevor-Roper[2]

In August 1941, Winston Churchill received a recommendation from Lord Cherwell—'Cherwell, when the war began, was plain professor Lindemann' as the doggerel goes—that the work that the British had started on the bomb should be supported. Churchill responded with 'Although personally I am quite content with the existing explosives, I feel we must not stand in the path of improvement'.

Frederick Lindemann was Churchill's close friend and advisor during World War II. He was a physics don at Christ Church College, Oxford. In 1941 he was elevated to a peerage and took the name Lord Cherwell after a stream that flowed near the college.

As I mentioned, on that early morning in 1957 in Nevada I also saw something that surprised me—the gangly figure of a fellow named Al Peaslee, who had been a graduate student at Harvard when I was. After taking his degree he vanished and now here he was in the desert. He had clearly been here before and he gave me some advice on how to view the explosion. Then he said something that I have never forgotten. Pointing to a small group of men standing in isolation he said, 'They're British and we don't tell them anything.'

I wanted to ask Peaslee what he meant. After all we and the British were allies in the Cold War. As an atomic bomb was about to be exploded I did not get the chance to ask him and I never saw him again. But over the years I tried to find an explanation. Now I have one in a book entitled *Churchill's Bomb: How the United States Overtook Britain in the First Nuclear Arms Race*, by Graham Farmelo [1].

In a sense this book is a dual biography of Churchill and Frederick Lindemann. Churchill's life story is well known, but Lindemann is probably unfamiliar. He was born in 1886 in Baden-Baden Germany to an American mother and a German father. He did his early scientific work on the continent but then relocated to England. His family was very wealthy, so, like Churchill, he was always surrounded by luxury. Unlike Churchill, he was a bachelor, vegetarian, non-smoker and teetotaler. He was an excellent athlete and competed in tennis at Wimbledon. He apparently met Churchill in 1921 when he partnered with Churchill's wife in an exhibition tennis match. By this time he was also famous for having in World War I produced a theory of a tail-spinning airplane which he tested by learning to fly. He was a good physicist, but never a great one, yet he got an appointment at Oxford which he maintained until his death in 1957. He was known both in and out of government as the 'prof'.

[2] H Trevor-Roper, unpublished.

Despite the fact that he was something of an anti-Semite, he was instrumental in relocating Jewish scientists to Britain in the 1930s. Among them was the German Rudolf Peierls and the Austrian Otto Frisch. Frisch had taken a walk in the woods during a 1938 Christmas vacation with his Aunt Lise Meitner in Sweden (where Meitner had taken refuge not long before). On the walk they analyzed some new results from her former Berlin colleague Otto Hahn and concluded that what Hahn had observed was nuclear fission. (There will be more about this in chapter 3.) The word soon spread like wildfire among physicists who realized its implications. But in February 1939 Niels Bohr published a brief note in which he pointed out that only the rare isotope of uranium—U-235—was fissile and since it was so hard to separate isotopes he later stated that it was unlikely that fission could be used in a nuclear weapon.

By this time Frisch and Peierls were together in Britain and as 'enemy aliens' they were not allowed to work on any classified subject, so they invented one. They asked themselves the question of what they could do with a supply of U-235 and how much would they need to make a bomb. Peierls worked out the theory of the critical mass and Frisch put in whatever numbers were available and came up with one kilogram as an answer. It was wrong by orders of magnitude. Fifty-two is about right. But the small number persuaded them that it must be taken seriously. They wrote up two memoranda.

The results were so sensitive that Peierls typed up the final document himself. By March it had found its way to Henry Tizard, who was well connected with Great Britain's wartime science establishment. He put together a meeting, and a new organization was created to work on something that was called 'Tube Alloys'. By 1941 a report had been written and a member of the committee—an outspoken physicist named Mark Oliphant—brought it to the United States with the mission of urging the Americans to get on with the bomb. Nothing much had happened since President Roosevelt had received a letter from Einstein two years earlier urging government involvement in nuclear energy. At this time the British knew more about nuclear weapons than anyone in the world. The British had a decision to make: should they make the bomb on their own or should they collaborate. Lindemann was strongly in favor of the former since he wanted assurance that the British would have the bomb after the war. But he was overruled and a somewhat uneasy collaboration began.

There is the matter of Klaus Fuchs, about whom we will hear a great deal in chapter 8. Fuchs was a German and although he was not Jewish he was driven out of the country for his anti-Nazi views. Eventually he found his way to Peierls and the two of them created the basic theory of isotope separation. A small group of British scientists, including Fuchs, were sent to New York to continue this work for the nascent Oak Ridge Laboratory. But in 1944 Peierls was sent to Los Alamos—'Site Y'. Fuchs was to follow, but at first he refused to go. This occasioned a letter on 14 July 1944 from James Chadwick, who headed the British delegation, to Peierls. The problem was that Fuchs wanted to go back to Britain. Chadwick wrote, 'I have now had a talk with Fuchs himself. He feels that he has a special contribution to make in England, whereas in Y he would be one of a number and can make no really significant difference.' Fuchs was persuaded to go to Los Alamos. What no one knew was that he was a Soviet spy and began transmitting information of the work at Los Alamos from the time he arrived. Some of this has been declassified by the Russians and you can see the whole plan for the plutonium device that was tested at Alamogordo and dropped on Nagasaki. It certainly saved the Soviets some years. Fuchs was unmasked in 1950. He had been working on the British nuclear program using information he had memorized at Los Alamos. I think that Peaslee's remark was a reflection on Fuchs. In 1957 we simply did not trust the British to keep secrets.

Churchill's last significant act in 1954, at the end of his second prime ministry, was to get approval for the British hydrogen bomb program. Their first one was tested in May 1957. The residue of this is four Vanguard submarines, each with sixteen Trident missiles, each of which can carry twelve hydrogen bomb warheads. The design is British, but it is based on an American prototype. The youngest Vanguard was commissioned in 1998. All need to be replaced and all seem to be pointless. When they function they wander around the ocean like lost souls. Is this something that Churchill really wanted? After Lindemann died in 1957 Churchill said, 'He is gone, and I am left to linger on.'

The first British bomb was tested in 1952, so in 1957 the British did not need us. It must be said that General Groves never trusted the British. The only British physicist he could abide was James Chadwick, the discoverer of the neutron. Groves knew what American intelligence had found out about the German program (see chapter 4) and he knew that the Germans had gotten nowhere and that the bomb would not be relevant in defeating them. It was the Soviets he was worried about. He expressed this at a dinner with Chadwick in Los Alamos in late 1944. Another guest was the Polish-born member of the British delegation, Joseph Rotblat. He promptly asked to leave Los Alamos and was granted permission, provided that he told no one where he had been. He had signed on to fight the Germans. He later became a noted advocate for nuclear disarmament, for which he won the Nobel Peace Prize. In 1946 the US Congress passed the Atomic Energy Act, which separated the British and American programs. It was in effect in 1957. It was amended the following year and after that the British and Americans conducted joint tests in Nevada. I imagine that Peaslee acted as host to his British colleagues.

Reference

[1] Farmelo G 2013 *Churchill's Bomb: How the United States Overtook Britain in the First Nuclear Arms Race* (New York: Basic)

IOP Publishing

One Physicist's Guide to Nuclear Weapons

A global perspective

Jeremy Bernstein

Chapter 3

The physics of fission

In the arsenals of the great powers, most nuclear weapons are no doubt hydrogen bombs. This means that a portion of their energy release (their 'yield') comes from fusion reactions of isotopes of hydrogen. But there are two reasons for me to discuss fission before fusion. First, all hydrogen bombs require a fission-bomb 'trigger', or 'primary'. Without a fission bomb you cannot make a hydrogen bomb. Second, even in a hydrogen bomb, most of the energy actually comes from fission. The fusion reactions, while they do produce energy, mainly have the function of enabling more fission.

Why does a fission bomb release so immense an amount of energy? I will give a paradoxical answer that will need a good deal of adumbration. My paradoxical answer is: because the atomic nucleus is so small. The nucleus is a roughly spherical chunk of very dense matter. The sphere is so small by everyday standards that a special unit of length has been introduced—the fermi, after the late Italian-American physicist Enrico Fermi. The fermi is a millionth of a millionth of a tenth of a centimeter. The radii of nuclei are of this general order of magnitude. The uranium nucleus, for example, has a radius of about seven-and-a-half fermis. How many of these tiny nuclei are there in a cubic centimeter of uranium? To get a grip on this huge number take a billion, square it, and then multiply that number by 10,000. Now you have a number that can be written as one followed by 22 zeros—10^{22}. That is about the number of uranium atoms in a cubic centimeter. Each of these atoms contains at its center a tiny and very dense nucleus. By now you may have an inkling of the plot. If we can find a nuclear reaction that generates even a minuscule amount of energy by household standards (although much larger than that from a chemical reaction), and if that reaction can be made to propagate, then because of the incredible number of nuclei in, say, a cubic centimeter, there will be an enormous amplification factor. Enter fission.

To understand fission we need a model of the nucleus. I have said that it is a very dense bit of matter of roughly spherical shape but minuscule size. Nuclear matter is

doi:10.1088/978-0-7503-1308-7CH3

made up of two kinds of particles—neutrons and protons. Protons carry a positive electric charge. Neutrons, which have no electric charge, have about the same mass as protons. In an atom, a cloud of negatively charged electrons, each with a mass of about a two thousandth of a neutron or proton mass, circulate at a great distance from the nucleus. If we were to enlarge the nucleus to be about the size of the Earth, then some of the electrons would be nearly halfway to the Sun. The electrons can go about their business without interference from the nucleus and vice versa. Atoms are electrically neutral, which means that the number of electrons is determined by the number of protons so as to balance the charge. To take two examples. The most common hydrogen nucleus consists of just one proton, so the atom has one electron. The uranium nucleus contains 92 protons, so the atom has 92 electrons. The business of these electrons is chemistry. In chemical reactions the atomic electrons of neighboring atoms interact with each other. For a given number of protons, there can be variants of the nuclei with different numbers of neutrons. These variants are called 'isotopes' and the isotopes all have sensibly the same chemical properties, since they have the same number of atomic electrons. I will give two relevant examples: hydrogen and uranium. The nucleus of ordinary hydrogen, as noted, consists of one proton. But the nucleus of 'heavy hydrogen'—the so-called 'deuteron'—consists of one proton and one neutron. There is also 'super heavy' hydrogen whose nucleus—the triton—consists of one proton and two neutrons. All three of these isotopes can combine with oxygen to produce water. There are nearly 30 known uranium isotopes. We will be concerned with only two—uranium-238 and uranium-235. Both of these isotopes contain 92 protons; the uranium-238 nucleus contains 146 neutrons, while uranium-235's contains 143 neutrons.

For the lighter nuclei such as the deuteron we can describe the behavior of the individual neutrons and protons. I wrote my PhD thesis on the deuteron, so I have a special feeling of affection towards it. But, when it comes to the heavy nuclei, there are so many particles that tracing them individually becomes a practical impossibility. We must resort to models of their collective behavior. One of the most successful, which was exploited by Niels Bohr in his analysis of fission, was what is called the 'liquid drop' model. What characterizes a liquid drop is that it has a definite shape that is maintained by a surface tension. This surface tension is caused by a strong nuclear force. It is so strong that it keeps the protons, which electrically repel each other, from flying apart. If this liquid drop is agitated it becomes deformed, the nearly spherical shape may become elongated and it begins to resemble a cigar. If it is elongated enough it splits apart into droplets and we have fission.

Put this way you might think that very smart physicists like Bohr and Hans Bethe, who back in the 1930s were completely familiar with the liquid drop model, would have predicted fission. But this is not at all what happened. After the neutron was discovered by the British physicist James Chadwick in 1932, it was clear to a number of people, especially Fermi, that as it had no electric charge the neutron would make an ideal nuclear probe, since it would not be repelled by the protons. Fermi headed a group of brilliant young physicists in Rome and they began a systematic investigation involving the bombardment of elements with neutrons. Fermi had discovered that, paradoxically, the slower the neutron was, the more readily it interacted.

He invented a wholly new discipline of slow neutron physics. In 1938 he was awarded the Nobel Prize partly for this and partly for nuclei he thought he had discovered, but had not. Using slow neutrons, one performs alchemy. After the neutron is absorbed by the target nucleus, new isotopes of different elements may be produced. Fermi was sure that when a uranium nucleus absorbed a neutron it would be transmuted into a new element that was heavier than uranium. Indeed he was so sure that he had discovered such elements that he gave them names: ausenium and hesperium. These are nice names. We now call them neptunium and plutonium, and Fermi did not discover them. What he had discovered, although he did not realize it, was fission.

This was 1934, and several other laboratories took up the study. The one that interests us is the Kaiser Wilhelm Institute in Berlin, together with its two radiochemists, Otto Hahn and Fritz Strassmann, and its physicist Lise Meitner. They repeated Fermi's experiment and also claimed to have discovered transuranic elements. This was 1938, and the roof collapsed on Meitner. Although she had converted to Christianity, as far as the Nazis were concerned, she was a Jew. She was Austrian, which immunized her until the Anschluss, when Austria and Germany were united. She was forced to leave Germany in a great hurry and ultimately ended up in Sweden. But she was able to keep some contact with Hahn and in December 1938 Hahn was informing her of something that he and Strassmann had observed and did not understand at all. They had been bombarding uranium with slow neutrons and one of the products was something that they could not distinguish from barium. Barium is an element more or less in the middle of the periodic table with 56 protons in its nucleus. Where did it come from?

It was now Christmas vacation and Meitner's nephew Otto Frisch came to visit her. He was a physicist who had also been driven out of Germany and he had sought refuge in Bohr's institute in Copenhagen. They met in the Swedish village of Kungälv near Göteborg. Frisch had brought his cross-country skis and she had brought Hahn's letter. They went for a walk in the woods, with Meitner trotting alongside her nephew. Frisch later wrote a charming account of that encounter. They sat down on a tree trunk and discovered fission. More exactly, they realized that the liquid drop model of the nucleus explained what Frisch a little later named fission. The important question was the energetics. You must have energy conserved in the process. How did this come about? The key lay in the masses of the nuclei: the uranium and the fission fragments. They knew that barium was one of the fragments and, by adding up the proton charges, they could deduce that the other was the inert gas krypton which has 36 protons in its nucleus. Thirty-six and 56 add up to 92, the number of protons in the uranium nucleus. I have always found it quite wonderful that the name of the planet on which Superman was born and the name of one of the elements produced in the first fission experiment are the same. Frisch says that his aunt knew the masses of these isotopes. Why does that matter? Here we invoke Einstein's equation $E=mc^2$.

It was known from the early 1930s how the binding energy per nucleon decreased by about 1 MeV as one went from barium to uranium. Considering the number of nucleons available, this would release about 200 Mev if uranium could be split into two barium-like nuclei. This information was available to everyone. What Meitner and Frisch did was to give a mechanism to the splitting using the liquid drop model of these nuclei. This is the energy released in the fission, which is mainly the energy carried away by the fission fragments. When you do this you get an answer that may surprise you. To understand this we must understand units of energy. I begin with something familiar: an electric light bulb. You all know that these bulbs are graded in watts. But you may not have realized that a watt is not a unit of energy. It is a unit of power—an amount of energy per second you take from the electric grid to illuminate the light bulb. If we take a watt and multiply it by a second we have a watt-second, which is a unit of energy. Your electric bill is usually measured in kilowatt-hours, which are also a unit of energy. Physicists have a name for a watt-second. We call it a 'joule', which is pronounced 'jewel' or 'jowl' depending on which historian of science you read. James Jewel or Jowl was a nineteenth-century British physicist—actually James Joule—who did important work on the conservation of energy. (In pronouncing his name I am reminded of what my former *New Yorker* colleague Burt Bernstein used to say when asked whether he pronounced his name Bernsteen or Bernstyne like his brother Leonard. Eether or Eythor, Bert always answered.) So the joule, however you pronounce it, is a watt-second. If you have a hundred-watt bulb and keep it on for an hour then you will have used 360,000 joules of energy.

Then how many joules are created in one act of uranium fission producing barium and krypton? The answer is 10^{-11}— the inverse of one followed by 11 zeros. This is an absurdly small amount of energy. If this was all there was to it there would be no bomb and no reactors and I would not be writing this book. Since I am, there must be something more to it.

Frisch returned to Copenhagen to find Bohr about to leave for the United States, where he was going to spend the spring semester in Princeton. He told Bohr what he and his aunt had done and Bohr's reaction was to say 'What idiots we have been.' How was it possible not to have thought of fission? Once in the United States the news of fission spread like wildfire in the physics community and many people began working on it. Bohr and the Princeton physics professor John Wheeler wrote a monumental paper on the liquid drop model of fission. In the course of this Bohr had an epiphany. He discovered the difference between an isotope that is 'fissionable' and one that is 'fissile'—to use the modern terminology. Take the case of uranium. If you take a sample of uranium from a mine over 99% of it will consist of the isotope uranium-238. Less than one per cent will be uranium-235, which has three fewer neutrons in its nucleus. What Bohr realized was that, due to a quirk in the nuclear physics, while uranium-238 is fissionable, uranium-235 is fissile. To fission uranium-238 the neutrons incident on it must have a minimum or threshold energy. A neutron with less energy will not fission the nucleus. On the other hand, uranium-235 is fissile—neutrons of any energy, no matter how small, will fission it. Bohr understood at once that creating a bomb using fission energy would require the separation of isotopes. Because these isotopes have the same chemistry he decided that this was hopelessly difficult, so, until the fall of 1943, when he escaped from Denmark to England and learned about the Manhattan project, he assumed that nuclear weapons were impossible. Henceforth when I discuss the fission of uranium I will mean the fission of uranium-235.

Bohr's epiphany by itself does not solve the problem of how to amplify the tiny energy produced in a single fission of uranium. I left something out. The fission fragments are accompanied by neutrons emitted in the fission process. Sometimes one neutron is emitted, sometimes two, sometimes three, and occasionally more. For uranium-235 the average is a little more than two, so to keep things simple I will round it off to two. I will also assume that each of these neutrons will in turn fission another uranium nucleus. In fact, the most probable thing is that the neutron will bounce off the next nucleus, but if the uranium sample is large enough, so the neutron does not escape through the surface, it will sooner or later fission a nucleus. Thus the first generation produces two neutrons, the second four, the third eight, and so on—a chain reaction. For reasons that I will now explain, I am interested in the 80th generation. This produces two to the power of 80 neutrons. But this is about equal to ten to the power of 24 neutrons. Why am I interested in this number? Because it is about the number of uranium nuclei in a kilogram of uranium. Thus 80 generations of fissions will produce enough neutrons to fission a kilogram of uranium-235. This is the amplification mechanism we have been looking for.

Uranium is very dense—about nineteen times as dense as water. A kilogram is a little over two pounds. If this amount of uranium were made into a sphere it would be about the size of a squash ball. Such a squash ball would weigh over two pounds. If we fissioned all the nuclei in it, how much energy would be generated? If we multiply the number of nuclei in the kilogram by the fission energy generated by each nucleus we arrive at the remarkable result that the total energy generated is equivalent to 10,000 tons of TNT. Think about that. The fission energy potentially available in a squash ball-sized lump of uranium is enough to flatten a city. However, the existence of this energy by itself does not guarantee an explosion. By definition an explosion is the production of energy in a very short time. We must investigate how long it would take to fission a kilogram of uranium. The key here is the speed of the neutrons produced. In fission, the heavy fission fragments take away most of the energy of motion—the kinetic energy—but the neutrons take away enough so that they are moving very fast—about a tenth of the speed of light. How far does a neutron have to go on average before it produces another fission? In uranium-235 it turns out that this is about 13.5 centimeters. The neutron can cover this distance in about ten nanoseconds—ten billionths of a second. This is such an important unit of time in this business that the practitioners have given it a name—a 'shake'—one shake of a lamb's tail. To fission a kilogram we need 80 generations, which takes eighty shakes—about a microsecond—a microsecond! This is very important to keep in mind. In a nuclear explosion all the initial action takes place in microseconds. Later it slows down to milliseconds. Unless you have a very high-speed camera you cannot record what happens.

The alert reader—and I know who you are—may have smelled a rodent here. You may object that the mean free path for fission is about 13.5 centimeters, but the diameter of a squash ball is only about four centimeters. Do not many of the neutrons leak out of the surface of the spherical kilogram, taking them out of the chain reaction? The answer to this question is yes, and it brings us to the matter of the critical size needed to just sustain the chain reaction. If we know the size then we can find the mass by multiplying the volume by the density. This was a major focus of the work at Los Alamos. If you are discussing a spherical shape then the geometry is simple enough so that you can make good theoretical predictions. But Los Alamos needed odd shapes and the critical masses had to be determined empirically. This is something that Feynman called 'tickling the tail of the sleeping dragon'. It was a very dangerous operation, adding bits of fissile material a little at a time until the configuration went critical. For uranium-235 the critical mass of a sphere is about 50 kilograms—a bit over a hundred pounds. A sphere of uranium with this mass is about the size of a baseball. This volume will have more nuclei in it than the squash ball. It might take another shake or two to fission them all, but that is not going to add much to the time.

This brings up something about the German wartime nuclear program, to which I devote the next chapter. In 1942, the official program was led by Werner Heisenberg. In June 1942 he went to see Albert Speer, the newly appointed Minister of Armaments and Munitions. Speer wanted to know how big a uranium bomb would be, by which I think he meant how much uranium would it require. Heisenberg told him it would be about the size of a pineapple. I have discussed this with German colleagues and pointed out to them that this must mean that Heisenberg never correctly calculated the critical mass of uranium. The Germans never produced any uranium-235, so they could not have measured it. One of my German colleagues noted that, in wartime, pineapples were unknown in Germany. I am sure that Heisenberg was enough of a man of the world to know that a pineapple is larger than a baseball.

A reprise: beyond imagining

Since nuclear testing has become 'sanitized', hidden from view, many people seem to have acquired the idea that nuclear weapons are a mere 'improvement'—somewhere on the continuum that begins with a firecracker, then a stick of dynamite, then the 5,000 pounds of high explosives packed in the rented Ryder truck that Timothy McVeigh used on 17 April 1995 to partially blow up the Murrah building in Oklahoma City, then a few notches over; Little Boy or Fat Man. I want first to persuade you that it is not a few notches over, and then, in chapter 5, to say more about how these things work. It is instructive to compare McVeigh's act of destruction with Hiroshima. I note that Little Boy would have fitted into Timothy McVeigh's Ryder truck. In addition to the damage that McVeigh's device caused to the building, it killed 168 people. The Hiroshima bomb—Little Boy—was equivalent in its explosive power to about 32 million pounds of TNT—32 million! It would take 6,000 Ryder trucks full of TNT to produce an equivalent explosion. The Hiroshima bomb destroyed the infrastructure of an entire city. By December 1945, at least 90,000 people had died. Near the epicenter of the explosion a wind was produced that was stronger than any known typhoon. Houses were turned into kindling and a firestorm resulted.

The only comparable bombing damage on Japan was done on 6 and 10 March 1945, when 334 B-29 bombers dropped two-million pounds—a kiloton—of incendiary bombs on Tokyo, killing perhaps 100,000 people. The one Ryder truck-sized Hiroshima bomb produced an explosion that was 15 times greater. We are clearly dealing with something very different here and I want you to understand why. In chapter 5 I say more about the science of these weapons.

IOP Publishing

One Physicist's Guide to Nuclear Weapons

A global perspective

Jeremy Bernstein

Chapter 4

Nazi Germany's failed program

Anyone who has studied the attempt to develop nuclear energy in Germany during the Second World War immediately confronts three questions.

1. Were the Germans trying to make a nuclear weapon?
2. If so, how close did they come?
3. If they had succeeded, would they have turned this weapon over to Hitler?

All three of these questions have been answered in very different ways by different people, but before giving my answers, and the alternatives, it is important to clarify what one means by 'the Germans'. Which Germans? To understand the issue let me review a little of the history. As I said in the previous chapter, in December 1938 the German physical chemists Otto Hahn and Fritz Strassmann made the first observation of the fissioning of atomic nuclei. They bombarded uranium with slow neutrons and noted that one of the end products seemed to be the relatively light nucleus barium. A month later the Austrian-born Jewish physicists Lise Meitner and her nephew Otto Frisch—who had taken refuge in Scandinavia after fleeing Germany, she to Sweden and he to Denmark—made the correct analysis of the Hahn–Strassmann observation, namely that the uranium nucleus had been split, barium being one product, with the release of energy. Clearly Meitner, who had been a colleague of Hahn's in Berlin, and Frisch, who had been in Hamburg, should not, when answering these questions, be counted among 'the Germans'. Nor should Rudolph Peierls, who had emigrated to Britain from Germany. It was the 1940 memorandum of Frisch, by that time also in Britain, and Peierls that persuaded the Allies that an atomic bomb was a real possibility. Frisch and Peierls realized that if you could separate—they said less than a kilogram; sixty kilograms is closer to the correct value—the rare uranium isotope U-235 from the common isotope U-238, you could generate an explosive nuclear reaction. Prior to this calculation it was thought that tons would be required—an amount so massive that scientists such as Niels Bohr had decided that nuclear weapons were a practical impossibility.

He thought no such quantity could realistically be produced by the separation of the uranium isotopes.

One also cannot count among 'the Germans' physicists such as the Nobelist James Franck, or Hans Bethe, or, needless to say, Albert Einstein, who were driven from Germany in the early 1930s. Nor can you count non-Germans such as John von Neumann and Eugene Wigner, both Hungarian Jews, who were beginning their careers in the German university system. They were forced to emigrate. And there were Italians such as Emilio Segrè and Enrico Fermi who chose to leave Italy because of the racial laws that were a German infection. But even inside Germany there were German scientists whom one cannot include among 'the Germans'. Strassmann himself, as an avid anti-Nazi, was deprived of his livelihood during the war. And then there was the remarkable Max von Laue, also a Nobelist, who publicly refused to deny Einstein credit for his relativity theory, something that most of the other German physicists were all too ready to do. It is a miracle that Laue did not end up in a concentration camp. There is also the notable case of Gustav Hertz. Hertz, who had Jewish ancestry, shared the 1925 Nobel Prize in physics with James Franck. His students and colleagues had such admiration for him that when he lost his university position they hid him in the Siemens industrial laboratories in Berlin during the entire war. He was also not one of 'the Germans'.

But there were also Germans whose racial pedigree was acceptable but who did not have the sort of academic prestige required for them to be taken entirely seriously. A case in point is Manfred von Ardenne. He was an inventor and a scientific entrepreneur of significant talent, although he had no academic qualifications in physics. He managed to persuade the German Post Office to sponsor work in nuclear physics, which was carried out on his estate in Berlin. In early 1941 one of his associates, Fritz Houtermans, who had had the unusual distinction of having been jailed by both the Soviet secret police and the Gestapo, made the observation that what came to be called plutonium was an even better nuclear explosive than uranium. The same discovery had been made independently by the more conventional German physicist C F von Weizsäcker and the American physicist Louis Turner a few months earlier. Von Ardenne's group also made significant progress in the separation of the uranium isotopes. While what they had done was more or less ignored by the German scientific establishment, it was important enough that when the Soviets occupied Berlin they shipped von Ardenne, his equipment, and his colleagues east, where they helped to make the first Soviet bomb. These people too should probably not be counted among 'the Germans'. Who, then, is left?

In the fall of 1939, German Army Ordnance decided that it was imperative to sponsor a program to study nuclear fission for use as a possible weapon. They took over the Kaiser Wilhelm Institute for Physics in Berlin and began drafting physicists and chemists to work on the project. The most important 'draftee' was Werner Heisenberg, who was certainly one of the greatest physicists of the twentieth century. It was Heisenberg who had the first truly quantum mechanical mind. He was to quantum theory what Einstein was to relativity. His prestige in Germany was enormous. In addition to Heisenberg, some sixty scientists from various institutions eventually joined the project. They came to call themselves the *Uranverein*—the

Uranium Club. Clearly these are 'the Germans' one wants to consider. But there was also an industrial infrastructure. To take one notorious example, the Degussa Company had taken over a metallurgic company called Auer since its previous owners were not Aryan enough. Degussa-Auer used slave labor—some 2,000 women from the concentration camp at Sachsenhausen—to produce uranium oxide for the *Uranverein.* These too should be counted as part of 'the Germans'. Degussa, incidentally, is still in business. It has a website with a smiling face. One of its activities had been supplying the Iraqis with nuclear material. We may now return to the three questions.

First, was the *Uranverein* trying to make a nuclear weapon? It is hard to see how there can be much of an argument about this, at least until early 1942, when the army withdrew from the project, which was then funded by the Reich Research Council. At that point most of the effort went into trying to make a functioning nuclear reactor. This enabled the Germans, and their spokesmen, to claim that the project was really about the peaceful use of nuclear energy. In 1958, the journalist Robert Jungk published *Brighter Than a Thousand Suns* [1] in which he stated that 'It seems paradoxical that German nuclear physicists, living under a saber-rattling dictatorship, obeyed the voice of conscience and attempted to prevent the construction of atomic bombs, while their professional colleagues in the democracies, who had no fear, with very few exceptions concentrated their whole energies on the production of the new weapon.' But, as Jungk himself came to realize, although many years later, once plutonium was discovered a reactor is a weapon since reactors are used to make plutonium, a fact that the *Uranverein* was very explicit about when they made appeals to the government for funding. There was also a smaller program to try to design some sort of exploding reactor—like a miniature Chernobyl—which could have spread radioactive material over a large area. So to the first question my answer is 'yes'.

Then how close were they to making a bomb? To this I would answer 'not very'. A variety of explanations have been offered, including the fact that Germany was being bombed and in any event did not have the necessary industrial capacity, that the *Uranverein* was not trying and even attempted to sabotage the project, and then my own favorite, which is incompetence. One must keep in mind that in early December 1942 Enrico Fermi with an infrastructure that was certainly no greater than that available to the *Uranverein* succeeded in making the first functioning nuclear reactor in an abandoned squash court at the University of Chicago. This was something the Germans never achieved. The difference, as far as I am concerned, is that our program had Fermi while the Germans had Heisenberg. Heisenberg was simply not a good engineer, something which his ego prevented him from acknowledging. If the Germans, who started first, had been able to create a self-sustaining chain reaction I think the whole project would have taken on a different sense of urgency. While they would never have had the industrial capacity to create an Oak Ridge—where our uranium was separated—they might have been able to make a Dimona—the Israeli plutonium-producing reactor which probably produces enough material for a few bombs a year.

This brings me to the last question: would they have turned the bomb over to Hitler? Here, our own experience may be relevant. Once the Manhattan Project to build the bomb was launched in August 1942 it became a project of the United

States Army under the command of Major General Leslie Groves. Men were drafted and sent as soldiers to Los Alamos. In the beginning there was even a discussion about giving the scientists simulated ranks and putting them into uniform. Oppenheimer liked the idea until cooler heads prevailed. When the bomb was built the army simply took possession of what it had bought and paid for. A few of the scientists tried to enter into the decision process of what to do with the bomb by organizing an action group, but they had no say in what finally happened. Can anyone imagine that it would have been any different in Germany? If the project had shown more signs of succeeding the army would have kept control of it and the bomb would have become the weapon of choice.

What is remarkable is that now, more than a half century after the fact, the activities and intentions of the *Uranverein* still provoke vivid and polemical debate. To take two examples. In 1993 the journalist Thomas Powers published his book *Heisenberg's War* [2] in which he tried to argue that Heisenberg deliberately sabotaged the German project by withholding knowledge about the bomb and even attempting to pass some information to the Allies. Powers's book unleashed a hailstorm of protest, the most recent example of which is *Heisenberg and the Nazi Atomic Bomb Project* by the Pennsylvania State historian Paul Rose [3]. The title is a clue to the tone. One may wonder why these matters have not been settled. After all, you are not dealing with the reconstruction of a prehistoric civilization from a few drawings on the wall of a cave. This was a project that created a paper trail of hundreds and hundreds of documents of all kinds. The principals were, at least until a few decades after the war, all alive and prepared to tell their stories. How then are such radical disagreements possible? I will give two examples, both suggested by Professor Rose's book, examples that will take us deeper into the subject and show just how difficult it is.

Early in his book Professor Rose quotes what he refers to as a 'bizarre letter' purportedly written in the spring of 1970—the letter is undated—from Heisenberg to a woman named Ruth Nanda Anshen. Anshen was the editor and guiding spirit of a series of books—*World Perspectives*—by outstanding thinkers including a book by Heisenberg and one by I I Rabi, both of whom were on her board of editors. This is the letter. It is taken directly from Anshen's book *Biography of an Idea* [4].

> Dear Dr [sic] Anshen:
>
> I have finished reading in your *Perspectives in Humanism* [sic] series the volume written by Professor Rabi entitled *Science: The Center of Culture.* I should like to review this important volume. However, I must say to you that I shall have to take exception to Dr Rabi's statement that 'such a tremendous undertaking as Oak Ridge, with huge, combined efforts of science, engineering, industry and the Army would have been impossible in bomb-ridden Germany.'
>
> Dr Hahn, Dr von Laue and I falsified the mathematics in order to avoid the development of the atom bomb by German scientists.
>
> Sincerely yours,
> Werner Heisenberg

When I read this letter, especially the last sentence, I did not think it was 'bizarre', I thought it was incredible. Heisenberg was always very careful not to make explicit claims like this. He let other people like Jungk and, posthumously, Powers and even his wife do it for him. Why would he make such a claim in such an apparently casual way? If he believed it why had he not said so in any of his published accounts of his wartime activities? Then there is the phrase 'falsified the mathematics'. What does that mean? The atomic bomb does not involve some mathematical equation like $E=mc^2$ that you can 'falsify'. It involves hundreds and hundreds of engineering details. And there is also the mention of Hahn and Laue as co-conspirators. Hahn was, in the first place, not a mathematician but a physical chemist. While he disliked the Nazis, he liked his comfort even more. There is no indication that he ever risked his life—which is what this would have meant—for anything except helping Meitner to escape Germany. He escorted her onto a midnight train that crossed the Dutch border at a little used point. Later she went to Sweden. But Meitner was so infuriated by his general laissez-faire attitude towards the Nazis that after the war she wrote him an exceedingly angry letter. And Laue, while he was at the Kaiser Wilhelm Institute had nothing to do with the *Uranverein.* In any event, he was not a nuclear physicist. The whole notion of his involvement in this is absurd. Heisenberg would surely have known that anyone with the slightest knowledge of these matters who read this claim about Laue and Hahn would find it totally crazy.

Rose notes that there is something else that is odd about this letter. Mrs Anshen gave her papers to Columbia University, including all her letters, but this one is missing. Rose remarks, 'Nevertheless, it seems beyond doubt that the letter was genuine; its quotation in the Anshen memoirs is direct, and the letter itself is so startling that it could scarcely have been invented.' I was so 'startled' that I decided to try to check it out further. I have a colleague named Cathryn Carson who has been doing work on this history for several years. She has studied much of the Heisenberg *nachlass,* including the correspondence, which is largely in the Heisenberg archive in Munich. I asked her if she had ever come across such a correspondence. She said that indeed she had and that she would request permission of Helmut Rechenberg, who is in charge of the archive, to send me copies. Rechenberg and I have had strong disagreements about Heisenberg's wartime activities, but I am pleased to report that despite these disagreements he gave me permission to see the letters. He also gave Powers permission to see them, since Powers quotes from one of them in his book. Whether Rose was given permission I do not know. There are in fact three letters.

The first letter is in English and is dated 30 May 1970. It is from Anshen to Heisenberg. In it she requests that Heisenberg write a review of Rabi's book, which appeared in her old series *World Perspectives,* for her new series, *Perspectives in Humanism.* In Heisenberg's alleged letter he does not even have the series right. On 19 June Heisenberg replied in a letter of a page and a half in German. In this letter, which is addressed to 'Frau Anshen' and not 'Doctor Anshen', he asks her to inform Rabi that he is somewhat reluctant to write this review since he has strong disagreements with Rabi's chapter 7, which is on the German atomic bomb project. He is especially upset that Rabi has accepted the arguments that were first put forward by Samuel Goudsmit in his book *Alsos* [5] that the Germans would have been only too happy to have turned the bomb over to Hitler if they had had the competence to build it. Goudsmit had led a mission called 'Alsos'—'grove' in Greek—to follow the advancing Allied armies into Germany and capture as many of the *Uranverein* as possible, along with their equipment. In the event, nine members of the 'club', including Heisenberg, Hahn and von Weizsäcker, were interned in Britain for six months in an estate near Cambridge named Farm Hall. Laue was also interned, but mainly because it was thought that he might be better off in Britain than in post-war Germany. As it happened, British intelligence wired Farm Hall so that everything the Germans said to each other was recorded. In February 1992 the British finally released these transcripts and they are the most vivid record we have of the real thoughts of the *Uranverein.* Goudsmit had had access to them before writing his book. However, Goudsmit got a certain number of things wrong, especially his statement that the *Uranverein* had not thought of plutonium. He and Heisenberg, and later von Weizsäcker, got into a bitter exchange of letters and articles, so it is not difficult to see why Heisenberg objected to Rabi's reliance on Goudsmit's book.

The rest of the letter to Anshen is vintage Heisenberg. He constructs some sort of murky parallel between the dilemma posed by handing an atomic bomb over to Hitler and that of using a hydrogen bomb to end the war in Vietnam, which was still going on. 'Rabi,' he writes, 'completely overlooks the fact that the German physicists had about the same kind of psychological attitude towards putting a bomb in Hitler's hands as many Americans have today about the possibility of ending the American war with North Vietnam by dropping a hydrogen bomb on Hanoi.' He ends the letter by asking Mrs Anshen to take these matters up with Rabi so that they can avoid a public dispute. There is no mention of Oak Ridge and the bombing of Germany. There is no mention of falsifying data and there is no mention of Hahn and Laue. On 9 July Mrs Anshen replied to Heisenberg saying that she had spoken to Rabi and had decided that it might be better if Heisenberg did not review his book after all. She notes that 'Professor Rabi would not wish to enter into a polemical discussion with so great a physicist as you are'. Knowing Rabi as I did, I can just see him concocting this phrase—'so great a physicist as you are'—with some glee. Rabi knew his man. He had offered Heisenberg a job at Columbia in 1939 when Heisenberg was on his last visit to America just before the war. Rabi told me that Heisenberg turned him down, explaining that he did not want to lose his tenure in the German university system. But what should we make of all of this? I know what I make of it. Unless and until someone shows me the original of the letter that Rose quotes, as far as I am concerned, it is a chimera.

The second example is more profound. I feel responsible for it. I was the one that first wrote about it and I now think that what I wrote was wrong, although both Powers and Rose continued to embrace it. I will explain. Beginning in November 1977, on and off for two years, I conducted a series of tape-recorded interviews with Hans Bethe, which ultimately led to a three-part *New Yorker* profile. Bethe was the head of the Theoretical Division at Los Alamos during the war and therefore in the inner circle of Oppenheimer's advisers. When we began discussing Los Alamos

I asked Bethe if while they were working on the bomb they had had intelligence about what the German nuclear project was doing. Such intelligence, if it existed, was a very closely held secret during the war. I do not think, for example, that the people at Los Alamos knew about the activities of the Austrian-born chemist Paul Rosbaud. He remained in Germany throughout the war and passed information about the activities of the *Uranverein* through the Norwegian underground to the British. But Bethe did have some important information. He told me the following which I will give in the version he first gave to me and the version that, indeed, appeared in the *New Yorker*. This is the version that I no longer believe to be entirely accurate.

Bethe reminded me that in September 1941 Heisenberg had come to Copenhagen, where he met Bohr. The Germans had occupied Denmark in 1940 so that Heisenberg was a German visiting an occupied country. Indeed, one of the most distasteful of Heisenberg's wartime activities was visiting occupied countries. In December 1943, for example, Heisenberg went to Krakow at the invitation of his brother's school classmate Hans Frank, who was then the Governor-General of Poland. One wonders what they talked about. Frank was then enthusiastically engaged in supervising the extermination of the Polish Jews. The ostensible reason for Heisenberg's visit to Copenhagen was to be part of a German delegation that included von Weizsäcker to a conference of astronomers organized by the so-called German Cultural Institute in Copenhagen, an outfit set up to distribute German propaganda. Bohr boycotted this conference and there was some question as to whether he would see Heisenberg at all. It should be remembered that while in Copenhagen in the late 1920s and early 1930s under Bohr's relentless questioning Heisenberg hammered out what is known as the 'Copenhagen interpretation' of quantum mechanics, which involves probabilities and uncertainties. This is what we still use and what we still teach our students. Bohr wanted to at least invite Heisenberg to dinner, but his wife Margrethe objected. She felt that the whole visit was 'hostile'. She never much liked Heisenberg, but Bohr managed to persuade her. After dinner he and Heisenberg had a private talk, the contents of which have become one of the most disputed aspects of Heisenberg's entire wartime activity. There is not even an agreement as to where it took place. Bethe told me that he thought it was in some 'back alleys' in Copenhagen. Heisenberg has it as having taken place in either the Tivoli gardens or the *Faelledpark*. Others say that it was in either Bohr's study at home or his office at his institute.

The content and purpose of the talk are even less sure. There seems to be no doubt that nuclear weapons came up. Bohr's son Aage, a Nobel Prize-winning physicist in his own right and his father's closest wartime confidant, wrote, 'In a private conversation with my father, Heisenberg brought up the question of the military applications of atomic energy. My father was very reticent and expressed his skepticism because of the great technical difficulties that had to be overcome, but he had the impression that Heisenberg thought that the new possibilities [perhaps an implicit reference to plutonium] could decide the outcome of the war if the war dragged on.' But then what happened, or did not happen, is one of the most mysterious aspects of the whole business. Bethe told me that Heisenberg gave Bohr a drawing of what purported to be the design of a German nuclear weapon. He said that later—I did not ask him how much later—the drawing was 'transmitted to us in Los Alamos'. I also did not ask what Bethe meant by 'transmitted'—how? He told me that he and Edward Teller were asked to analyze the drawing and saw at once that it was a drawing of a nuclear reactor. Bethe said to me, 'But our conclusion was, when seeing it, these Germans are totally crazy. Do they want to throw a reactor down on London?' My *New Yorker* article, which included this quote, was the first time anyone had ever mentioned such a drawing in print.

Both Rose and Powers have accepted this version of events, but they have drawn almost opposite conclusions from it. Powers is convinced that Heisenberg committed the act of a traitor—a traitor to Germany—in that he passed classified information to Bohr. 'With this simple piece of paper Heisenberg had put his life in jeopardy.' In other words, Heisenberg was not a Nazi collaborator but a Resistance hero. For Rose, showing Bohr this drawing was an act of intimidation, to convince Bohr that the atomic bomb would play a role in the forthcoming 'Pax Nazica' (Rose's term). In short, Bohr should now collaborate with the German scientists because he would soon have no other choice. Both Powers and Rose have persuaded themselves that Heisenberg really handed Bohr a secret sketch of a German nuclear weapon of some sort. But did he?

In 1994 I was paying one of my periodic visits to the Rockefeller University when Abraham Pais, Bohr's biographer, called me into his office. He had received a phone call from Powers some months earlier during which Powers asked him what he knew about the drawing. Powers had been taken aback when he had received a letter from Aage Bohr which stated flatly that, 'Heisenberg certainly drew no sketch of a reactor during his visit in 1941. The operation of a reactor was not discussed at all.' Nor was plutonium, at least explicitly, since Bohr knew nothing about it until he was briefed about the Allied project in the fall of 1943 after he had escaped from Denmark. One would have thought that if Heisenberg had really wanted to communicate something important about the German program to Bohr it would have been plutonium. After my talk with Pais I was taken aback. Indeed I felt responsible for propagating a possibly serious error and I decided to try to find out what happened. But how?

The first thing I did was to write to everyone I knew who had had a senior role at Los Alamos to see if anyone had heard of this drawing. Bethe of course had and he repeated in writing what he had told me several years earlier. No one else seemed to know about it until I contacted Robert Serber. He turned out to be a gold mine. Serber was one of Oppenheimer's closest collaborators at Los Alamos. He also had a sense of history, so he kept important documents, and he had an excellent memory. He told me what had happened and he also sent me some corroborating documents. When Bohr arrived in Britain in September 1943 he was briefed by the British on the progress in the nuclear weapons program. Whether he told them what he knew about the German program we do not know. We do know that when he arrived in the United States in early December of that year he met with General Groves and, it appears, showed him some kind of drawing. I do not know if there is anything in Groves's papers that documents this. But I do know that Groves was sufficiently alarmed by what Bohr told him that he alerted Oppenheimer. On 31 December, just after Bohr arrived at Los Alamos with his son Aage, Oppenheimer called a meeting of a select group of staff members. Serber gave me a copy of the letter that Oppenheimer sent to Groves after the meeting, which lists the attendees. I managed to contact all of those who were still alive. Aage Bohr was one, as were Serber, Teller and Bethe. Serber recalled coming into the meeting a little late and that Oppenheimer said to him that they were discussing a proposal for a nuclear weapon that he said was Heisenberg's. He showed Serber the drawing and Serber also realized that it was a reactor. No one I contacted could say whether it was claimed that this was a drawing made by Heisenberg or a drawing made by Bohr from memory. The drawing itself seems to have vanished. Bethe and Teller then wrote a report of which Serber gave me a copy. They showed that this reactor could never explode like a nuclear weapon. No reactor can. This is what Frisch and Peierls had understood in 1940. A reactor operates with slow neutrons that have an enhanced probability of causing uranium or plutonium to fission. That is why the neutrons in a reactor must be slowed down or 'moderated'. The uranium in a conventional reactor is only slightly enriched with the U-235 isotope. A bomb on the other hand is made of nearly pure U-235 or plutonium, which are fissionable by fast neutrons. The entire explosive reaction in a bomb lasts only about a microsecond. Whether Heisenberg ever really understood this distinction is again one of the things that is vehemently debated. He said he did and Powers agrees with him. Goudsmit and Rose say he did not and I agree with them. I base my position on an analysis of the Farm Hall transcripts. The Germans were recorded just after they first learned about Hiroshima. It is clear to me that they did not have the most rudimentary understanding of how a nuclear weapon worked. In a few days Heisenberg figured it out and gave his fellow detainees a lecture about it. The comments at this lecture make it clear to me that the Germans were hearing about this for the first time.

When I read the report that Bethe and Teller wrote I realized that it was not any reactor that they were analyzing but a particular design that Heisenberg clung to even though it had been shown by other more junior theorists in the *Uranverein* that it was inefficient. Specifically, this design involved layers of uranium metal submerged in 'heavy water'. This is water that consists of oxygen and heavy hydrogen, a rare isotope with one extra neutron. The Germans used heavy water to try to moderate the neutrons in their reactor designs. But it was very hard to come by and the Germans never had enough. Fermi's reactor had lumps of uranium embedded in a purified graphite moderator. To me this meant that the drawing was something that came out of the *Uranverein.*

How can one reconcile this with Aage Bohr's absolute certainty—which he conveyed to me once again in a message delivered by Pais—that no reactor was discussed when his father met Heisenberg in 1941 and that no drawing changed hands. Here I will hazard a guess. I feel quite certain, unlike Rose and Powers, that Heisenberg never gave Bohr this drawing. It simply does not fit Heisenberg's character. He was never a Nazi, but he was a patriotic German. He told several people both during and immediately after the war that he had wanted the Germans to win it. He told Bethe that he wanted the Germans to win because if the Allies won he was sure they would level Germany and destroy German culture. He was sure, he told Bethe, that once the Germans won the war the 'good Germans' would take over and restore things to the way they were before the Nazis. What I think happened is that someone from the *Uranverein* visited Bohr and gave him this information. A likely candidate, it seems to me, is the *Uranverein* physicist Hans Jensen who visited Copenhagen in 1942 and did discuss the German program with Bohr. Bohr was still persuaded that in any practical sense nuclear weapons were impossible, so he probably filed Jensen's report somewhere in his head and when he was briefed in Britain he recalled what he had been told. I cannot prove this, but neither can Powers or Rose prove their claim. It may be that there are some things about this history that we will never know for sure.

When I first saw the Farm Hall transcripts in the spring of 1992 I thought that they would make a great play if one could make the references and the physics clear. Unfortunately, from the historical point of view, the surviving transcripts cover only a minute part of the conversations that were actually recorded. The listeners, including General Groves who was being sent them in Washington as soon as they were prepared, were interested in knowing what the Germans knew about nuclear weapons and whether they were likely to defect to the east. One of the copies that I have has annotations by Groves. Every time the Soviet Union is mentioned there is a seismic reaction—underlining and exclamation points. He was seemingly not very interested in Hahn's almost suicidal reaction to learning that his discovery of fission had been used to destroy Hiroshima. Nor was he interested in the surreal award of the Nobel Prize in November 1945 to Hahn—and Hahn alone—for the discovery of fission, with the award being made three months after Hiroshima, when Hahn was still in detention. What the Swedish Royal Academy could have been thinking of I cannot imagine and why Strassmann, Frisch and Meitner were not included I cannot imagine either.

What a pity that we do not have a recording of what really went on between Heisenberg and Bohr in September 1941. One is free to let one's imagination wander. This is just what Michael Frayn did in his play, *Copenhagen* [6]. Frayn is not a physicist, but he has evidently read a great deal about this. Some of it he got wrong. Hahn did not fission uranium into two barium nuclei, but rather one barium and one krypton. The Soviets at Bohr's institute when Heisenberg was there were not Gamow and 'Landers', but George Gamow and Lev Landau, who was arguably the most brilliant Soviet physicist of his era and later ran afoul of Stalin. These are minor errors. What is not minor is a bit of dialogue he has given to Bohr. Referring to Heisenberg, he says, 'A White Jew. That's what the Nazis called him. He taught so-called Jewish physics and refused to stop. He stuck with Einstein and relativity, in spite of the most terrible attacks.' It is true that Heisenberg, who was, like most Germans, indeed like most of the *Uranverein,* not a member of the Nazi Party, was attacked by the Nazis for having done 'Jewish physics'. He was called a 'White Jew'—what a strange epithet. But in July 1938, with the help of a family connection, he was vetted by Heinrich Himmler himself, who warned him to disassociate himself from the 'personal and political attitude of the scientists involved'. In other words, you could have relativity but not Einstein. This was codified in a conference in Seefeld in the Tyrolean Alps in November 1942 attended by 30 German scientists, including Heisenberg and von Weizsäcker, who wrote the summary report which stated that 'At the Seefeld meeting the opinion was expressed however that one must reject the imposition of the physical relativity theory into a world philosophy of relativism, as has been attempted by the Jewish propaganda press of the previous era.' The report also said that 'Einstein merely followed up the already existing ideas consistently and added the cornerstone.' This does not sound like sticking to Einstein and relativity to me. Nonetheless I think Frayn captured Heisenberg's moral ambiguity. He resisted the temptation of having Heisenberg pass the drawing on to Bohr. His Heisenberg is neither a Resistance hero nor a simple Nazi collaborator, but something more interesting and, indeed, more troubling. Frayn raises the question of why Heisenberg—and Bohr for that matter—never did the relatively simple calculation that Frisch and Peierls did that showed that a bomb could be built. Certainly either of them could have done it. The suggestion of Frayn's play—and this may be part of the truth as well—is that somewhere deep in their psyches they were held back because they did not want to know the answer.

A little added note. Some years ago I went with Frayn to visit Farm Hall. From the outside it looked like a modest estate. But the interior revealed a real manor house. It was then owned by a well-to-do Cambridge don. When he bought it he knew nothing of its wartime history. He did receive from time to time visits from a few of the former inmates who had come back with their families for sentimental reasons. But the real revelation came when he had to replace a floor. Underneath the floorboards was a maze of wires. These had been connected to the listening devices used to spy on the Germans. The wires were still there.

References

[1] Jungk R 1958 *Brighter Than a Thousand Suns* (New York: Harcourt Brace)

[2] Powers T 1993 *Heisenberg's War* (New York: Da Capo Press)

[3] Rose P 1998 *Heisenberg and the Nazi Atomic Bomb Project*, 1939–1945 (Oakland, CA: University of California Press)

[4] Anshen R N 1986 *Biography of an Idea* (Wakefield, RI: Moyer Bell Limited)

[5] Goudsmit S 1996 *Alsos* (New York: AIP Press)

[6] Frayn M 1998 *Copenhagen* (New York: Anchor Books)

IOP Publishing

One Physicist's Guide to Nuclear Weapons
A global perspective
Jeremy Bernstein

Chapter 5

More bomb physics

Imagine that you have acquired a hundred pounds or so of highly enriched uranium, how would you make a bomb? The first thing that might occur to you is to make two hemispheres, each of which has less than the critical mass, and then bang them together. This brings up an important point. The critical size of a sphere of uranium is just that size at which the number of neutrons escaping through the surface is equal to the number being created by fissions in the interior. This equilibrium is not going to produce an explosion. We need a super-critical mass so that more neutrons are being produced than are escaping. Banging two sub-critical hemispheres together will do this, but we can actually do better. This involves something that is essential to understand for the plutonium bomb. The figure we gave for the critical size supposed that the uranium is at its normal density. Suppose, however, that we take our sphere and squeeze it so that it remains a sphere of reduced volume but with the same amount of material. This will increase the density. At this increased density the distance between the nuclei is shorter and therefore the distance between fissions is shorter. This means that the critical size and hence the critical mass is smaller. In fact the critical mass falls off as the inverse square of the density. If we double the density the critical mass is reduced by a factor of four. Conversely, if we reduce the density by a factor of two the critical mass increases by a factor of four. This is what shuts the chain reaction off in the bomb. The immense heat very rapidly expands the uranium, which has become a gas (more exactly, a plasma). Neutrons can no longer find other uranium nuclei to fission and the chain reaction stops. This happens so fast that most of the uranium in the bomb is never fissioned. In Little Boy, which did bang two chunks of uranium together, only two per cent was fissioned. The rest wafted off into space.

In designing Little Boy, the observation about the density was used in the following clever way. Suppose we start with a solid sphere of (highly enriched) uranium and drill out a sizeable cylindrical hole—one into which we can fire an artillery shell. This way we have lowered the density of the remaining object because part of it has no mass and therefore zero density. Thus we have increased the critical mass of this structure and we can add additional uranium without the thing going

doi:10.1088/978-0-7503-1308-7ch5

critical. If it was at the original density it would actually be super-critical. Then, if we make the artillery shell out of uranium at the normal density, but sub-critical, we can fire it into the hole and produce a super-critical assembly. This is called the 'gun assembly' design. With a somewhat different geometry, it became the bomb. As I have mentioned, it was considered to be so simple to do that it was never tested before it was used. An entity that managed to get hold of this much highly enriched uranium, but had limited technical capacity, would probably try to design a nuclear device this way.

Starting with a sub-critical sphere and imploding it to produce a sphere with a smaller volume and therefore a higher density and a smaller critical mass is technically much more difficult. But if you can do it, since you do not lose any of the uranium in the implosion, what was sub-critical before the implosion is now super-critical after it. However, the technical problems in making the implosion are very substantial. In the gadget, one began with a solid sphere of plutonium—the 'pit'—wrapped carefully with machined explosive lenses around it, and then set off the explosives simultaneously—within nanoseconds of each other—at several points on the sphere.

Getting this to work was one of the hardest jobs done at Los Alamos. Below I explain why it was necessary. Fuchs gave most of the details to the Soviets who, as I have said, gave it to the Chinese. But the Chinese surprised everyone when, on 16 October 1964, in the desert at Lop Nor in the Uygur Autonomous Region in northwestern China, they successfully tested what turned out to be a uranium implosion device, demonstrating that they had a very sophisticated weapons program which included the separation of the uranium isotopes. We thought that they would use plutonium.

Plutonium: a remarkable element

Glenn Seaborg, Joseph W Kennedy, Edwin McMillan and Arthur Wahl discovered element 94 in Berkeley in December 1940. McMillan and Philip Abelson had discovered element 93 earlier that year. When Martin Heinrich Klaproth isolated element 92 in pitchblende in 1789, he called it uranium after the recently discovered planet Uranus. The scientists at Berkeley named elements 93 and 94 after the farther planets Neptune and Pluto. The discovery of plutonium was kept secret until after the war. At Los Alamos it was called '49'. This did not help much since Klaus Fuchs gave all the details about the bomb to the Soviets. The Germans also realized the value of element 94 for making bombs but they never could make a reactor to produce the stuff.

Seaborg's team bombarded uranium-238 with heavy hydrogen nuclei, or deuterons (a proton and a neutron; ordinary hydrogen has only a proton in the nucleus). This produces neptunium-238, which decays into the non-fissile isotope plutonium-238. Like other plutonium isotopes, Pu-238 decays into an alpha particle (two protons and two neutrons, identical to a helium nucleus) and an isotope of uranium. The half-life of this decay is a relatively short 87.7 years. (The fissile isotope Pu-239, created in a different way, has a half-life of more than 24,000 years.) It generates a good deal of heat: about half a watt per gram. The alpha particle is not a penetrating form of radiation, unlike beta particles (electrons or positrons) or gamma rays. It can be stopped by a sheet of paper. On a visit to the test site in Nevada in 1957, as I mentioned in chapter 1, I was given the pit of a plutonium bomb to hold. It was warm to the touch because of the alpha decay, but I was assured that I would not suffer any permanent damage. I was also told not to drop it. The pit was the size and weight of a bowling ball. This should have tipped me off that it was hollow since I never could have lifted a solid sphere of plutonium that size. The pits were being used to test 'boosted' weapons that use both fission and fusion.

All of the 20 known plutonium isotopes are unstable and, assuming they were about as plentiful as uranium at the time of the Earth's formation, in the four-and-a-half billion years since, they have decayed away. The uranium isotopes are also unstable, but uranium-238 has a lifetime comparable to the age of the Earth. Uranium-235 has a much shorter lifetime, which is why there is so much less of it now. Plutonium has to be manufactured. Glenn Seaborg and his collaborators confirmed its discovery in February 1941. Their Berkeley cyclotron created a secondary beam of neutrons to bombard uranium. This was a version of the experiment which had been tried since Fermi, but Seaborg had a more powerful neutron source. He also knew about fission, so he could weed out fission events from the nuclear transformations of uranium that lead to plutonium. He immediately realized that Bohr's argument that showed that uranium-235 was fissile applied, *mutatis mutandis*, to plutonium-239. In fact, plutonium-239 is even more fissile than uranium-235. The critical mass situation for plutonium is a little tricky, because it exists in different phases, called allotropes, with different densities. For the so-called delta phase that is used for making nuclear weapons, the critical mass of a bare sphere of plutonium is about 16 kg, a lot less than for uranium-235. Plutonium is manufactured in nuclear reactors. The plutonium for the gadget and for Fat Man, the Nagasaki bomb, was manufactured in the specially built reactors in Hanford, Washington on the Columbia River. The cleaning up of this site, which goes on to the present day, is a nightmare.

When the first significant quantities of reactor plutonium arrived at Los Alamos in the spring of 1944, it was clear that there was a serious and perhaps fatal problem. Heavy elements like plutonium and uranium have so many protons that repel each other electrically that they are on a knife-edge for fissioning. We have seen that colliding neutrons can fission them. But these elements also fission spontaneously—with no outside help. This is a potential disaster in weapons design because if there is a substantial amount of spontaneous fission before a super-critical mass is assembled, the chain reaction will be started prematurely and one will produce what the weapons designers call a 'fizzle'. These fizzles can involve explosions of hundreds of tons of TNT equivalent, but they are not a full-scale nuclear explosion. Spontaneous fission of uranium is manageable, but in the production of plutonium in a reactor both plutonium-239 and plutonium-240 are produced. Plutonium-240 has a high spontaneous fission rate. It is so high that the gun-assembly design is ruled out. The assembly is too slow and such a device would almost certainly go critical before it was fully assembled. At Los Alamos some thought was given to trying to separate these isotopes. But this would be even harder than separating the uranium isotopes, since the masses are even closer together.

On 4 July 1944 at an all-lab meeting Oppenheimer decided that the gun-assembly method could not be used for plutonium. Another method had been suggested as early as 1942 by the physicist Richard Tolman. It was adumbrated in a memo written with Robert Serber and one can find mention of it in Serber's 'Primer'—the lectures he gave at Los Alamos in April 1943. This was 'implosion', which I have mentioned before. Thus Oppenheimer turned the whole laboratory around to work on implosion. This is technically very complex, so it was decided that a test was needed and hence Trinity—and then Nagasaki with the same design.

I mention in passing that one isotope of plutonium, Pu-238, can be used to generate electricity safely. One of the first applications designed at Los Alamos in the 1960s was for use in pacemakers. A recent application is in the batteries powering the New Horizons space probe which flew past Pluto in 2015. The batteries were designed at the Idaho National Laboratory and are said to have cost 100 million US dollars. They contain about 10.9 kg of Pu-238 oxide pellets. This is a large fraction of what was available. Pu-238 is a rare commodity. The United States stopped making it in 1988 and was buying it from the Russians until they ran out. There is now a project to make more Pu-238 in the US, at a cost of tens of millions of dollars, by the neutron irradiation of element 95, americium, which is produced in small amounts in nuclear reactors. NASA does not have much left. A good deal of what they had is now heading for outer space. There is something fitting about this. Plutonium, which does not exist naturally, is created from uranium, which came to us in the first place from the collapsing of stars.

Implosion

The basic idea of implosion is very simple. Imagine a sphere of plutonium with a mass less than the critical mass so it does not explode when it absorbs a neutron. The critical mass is the highest mass which does not produce a runaway chain reaction when it absorbs a neutron. For plutonium-239 arranged as a solid sphere this mass is about ten kilograms, or 22 pounds. Plutonium is so dense that this corresponds to a sphere only about four inches in diameter. Now imagine that you construct a sphere of plutonium which has a mass that is less than the critical mass at the normal density. I stress this because in the case of a sphere if you increase the density the critical mass is reduced by the inverse square of the density increase. If, for example, you double the density, the critical mass is reduced to a quarter of its normal value. This process turns the subcritical sphere into an explosive super-critical sphere. It is important that in this process you preserve the spherical shape, because otherwise the results are quite unpredictable. The problem is to make this work in practice. This turned out to be very difficult.

From the beginning there had been a small group at Los Alamos headed by the physicist Seth Neddermeyer that worked on implosion without making a great deal of progress. When it was decided to focus the laboratory on implosion Neddermeyer was replaced by George Kistiakowsky, a Harvard chemistry professor of Ukrainian origin who was regarded as one of the world's greatest experts on explosives. Ordinary chemical explosives are needed to produce the compression. Experiments were begun and the results were awful. The final shapes were a tangled hash. Among other things, hollow spherical shells—not made of plutonium—were tried. At this point one of Oppenheimer's former graduate students, Robert Christy, made some calculations that showed that if solid spheres were used the results would be better. The device using this idea became known as the 'Christy gadget'. To produce the compression something that was known as an explosive lens was introduced. The idea came with the British delegation, but it was perfected by the Hungarian-born mathematician John von Neumann, who also had a degree in chemical engineering. Figure 5.1 gives a schematic drawing and below that is a schematic for the Christy gadget.

The way in which an optical lens works depends on the fact that light in air moves faster than light in glass. When a light beam impinges on a glass lens parallel rays will be bent towards each other and focus at a point behind the lens. The lens serves to concentrate the rays. Likewise you can wrap a fast explosive around a slow explosive and concentrate the detonation waves. This is easy to say but very difficult to realize in practice. Kistiakowsky and his group tested thousands and thousands of designs before they hit on one that would serve the purpose. Some features of this design were stolen by the spy David Greenglass and sent to the Soviet Union. These lenses must be arrayed around the plutonium 'pit' and nearly simultaneously detonated so as to preserve the spherical shape during the compression. The Christy gadget had 32 of these lenses.

This raises the crucial question of how these 32 lenses are detonated. It must be nearly simultaneous, for otherwise the compression will not be symmetric. For this purpose an entirely new detonator was invented—the 'exploding wire' detonator. This consists of a wire segment through which a very high current is run. This vaporizes the wire, causing a high temperature and a small shock wave that ignites

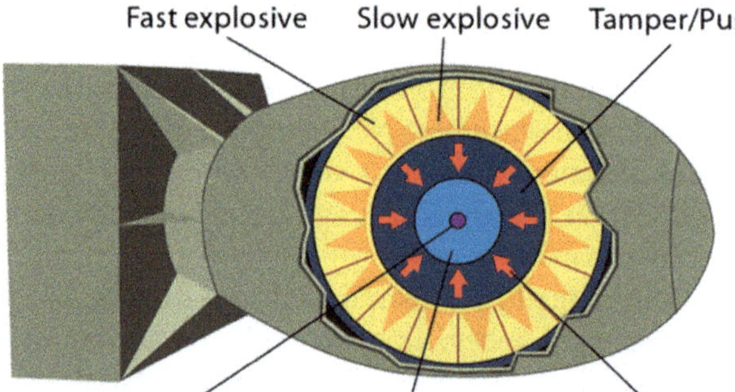

Figure 5.1. Implosion design. Credit: Wikimedia Commons. This image has been obtained by the author(s) from the Wikimedia website https://commons.wikimedia.org/wiki/File:Implosion_Nuclear_weapon.png, where it is stated to have been released into the public domain. It is included within this article on that basis.

the main charge. This detonator is very fast. All 32 detonators on the Christy gadget can ignite within about a microsecond. The implosion takes about a microsecond, which renders the spontaneous fission moot. One is impressed again and again by the scientific ingenuity that went into designing an instrument of death.

On 1 March 1945 Oppenheimer created the 'cow puncher' committee which was to ride herd on the fabrication of the device that was going to be tested in July. The basic design was frozen and it was transmitted to the Soviets in detail by Klaus Fuchs, who was said to know almost as much about the project as Oppenheimer. The Russians have released his report and I am continually amazed by how detailed it is. It begins by noting that 'The conducting of the first explosion on the atomic bomb is expected in July of this year.' Indeed, on the morning of 16 July the Christy gadget was successfully tested in New Mexico. A bomb of the same design was dropped on Nagasaki on 9 August.

* * * * *

What about the hydrogen bomb? In discussing the hydrogen bomb there is a serious limitation from security classification. Most of the details of fission bombs like the gadget are available in the open literature. You can, for example, find on the web what Fuchs gave the Soviets, which has been declassified by the Russian Federation. On the other hand, you cannot find the details of the hydrogen bomb—at least details that are assuredly correct—anywhere. That is a good thing. We do not want to make it any easier for countries like Iran or North Korea to make a hydrogen bomb. But I can tell you a few general principles. Hydrogen bombs, more correctly described as thermonuclear weapons, make use of nuclear fusion instead of fission (or, more accurately, as I mentioned before, in addition to fission). In fission, energy is obtained because a heavy nucleus is split and the products are less massive than the original nucleus—the mass difference producing an energy because of Einstein's equation $E{=}mc^2$. Fusion is the exact opposite. Two light nuclei fuse into a new nucleus or nuclei. What is paradoxical here at first sight is that while the nuclei that are the products of fusion have the same numbers of neutrons and protons as the fusing nuclei, they are less massive. This is what produces the energy. Let us begin with the simplest example—one neutron and one proton. When the neutron comes close enough to the proton it can be captured, producing a nucleus of heavy hydrogen—the deuteron. Suppose we want to get the free neutron and proton back again. We can do this by allowing a very energetic radiation quantum to interact with the deuteron. This is called the photo-disintegration of the deuteron. We will then discover that there is a minimum energy the quantum needs to make this process work. That is because the free neutron and proton are together more massive than the deuteron. In the reverse capture process a radiation quantum of the same energy is produced. This energy is given by Einstein's relation $E = mc^2$ using the mass difference between the free neutron and proton and the deuteron. This is where fusion energy comes from.

We might think that all we have to do to exploit fusion energy is to collect neutrons and protons in a container and let them fuse into deuterons. Were it that simple our energy problems would have been solved long ago. The difficulty is that the free neutron is unstable. It decays into a proton with a lifetime of about 15 min. Neutrons cannot be stored. We have to do the next best thing and go to isotopes of hydrogen where the neutron is bound and cannot decay in the same way. We could take two deuterons or, as it turns out, even better, one deuteron and one triton—the nucleus of super-heavy hydrogen with one proton and two neutrons. These two objects can fuse into a helium nucleus and an energetic neutron. The total energy released in this reaction is less than the energy released in a typical fission reaction. But the fusing elements are so much lighter that we make up for this since so much less mass is used in each fusion. There is no critical mass for this reaction and no theoretical limit to how much of the stuff we can use. But there is a catch. There is always a catch. Both of these nuclei have a net positive charge. Like charges repel, so these nuclei have a difficult time getting close enough to each other to fuse. In fact, in the world of classical physics they would essentially never come close enough and that would be the end of the story. But in the quantum world there is a new possibility. The nuclei can tunnel through the repulsive barrier. For this to happen frequently enough the nuclei must be very energetic, which means they must be in a very high-temperature environment. There is one place we know that this works and that is the Sun—and indeed all the stars. The central temperature of the Sun is about 13.5 million degrees Celsius. To reach a temperature at which the thermonuclear reactions could start, the force of gravity compressed the Sun. Then the reactions can proceed until the nuclear fuel is used up, whence the gravitational contraction begins again.

On Earth, we know one device that can produce temperatures like this—a fission bomb. This was understood by people like Hans Bethe and Teller in 1942. Even before there was a Los Alamos they began thinking about a 'super-bomb'. The first thought that might occur to you is to take a container of fusible elements and put a bomb in the middle of it and let it go off—something like dropping a match into a container of gasoline. This configuration, with all its innumerable adumbrations, is known as the 'classical Super'. None of the proposed designs seemed to work. While there was enough energy to ignite the fusion reactions, they would not propagate. The fusible elements cooled down too quickly. It was like trying to light a log with a match. Here is where things stood until 1951. At that time, the Polish-American mathematician Stanislaw Ulam made a breakthrough. Ulam had been at Los Alamos during the war, and fell in love with New Mexico, and decided to stay. He was a pure mathematician by training and temperament but, when the occasion arose, he could do very significant physics. He was one of the people who kept killing Teller's classical Super designs. I knew him quite well and used to tease him by calling him the 'mother-in-law of the hydrogen bomb', Teller supposedly being the father. Ulam's idea was to use a staged device. The first stage would be the fission bomb. The fusible material would be kept in a capsule separated from the fission bomb. His first idea was to use the material from the fission bomb to violently implode the capsule. The material in the interior would then be heated throughout by the compression. Teller made the important observation that the same effect was possible if one used the x-ray radiation from the bomb, which actually carries off most of the energy. These x-rays move with the speed of light, which expedites the process. It is the details of how this is done that are classified. But it is known that inside the capsule of fusible elements a hollow rod of plutonium is placed. The heated material now implodes the plutonium rod and fission is initiated. The energy from this fission ignites the fusion reactions. The fusion reactions produce energy but, more importantly, they produce high-energy neutrons. If one makes the capsule out of ordinary uranium, mostly U-238, then these neutrons cause intense fission. It is the fission that provides most of the energy produced by the device. I never cease to marvel at the ingenuity that went into creating a device for killing people.

The yield of hydrogen bombs can be about a thousand times greater than that of the Hiroshima or Nagasaki bombs. We are talking yields of millions of tons of TNT equivalent—megatons—600,000 Timothy McVeigh Ryder trucks. Before President Truman, on 31 January 1950, ordered a crash program to make a hydrogen bomb, and before anyone knew how to make one, there were debates among the physicists as to whether such a device should even be made at all. Oppenheimer was opposed. One of his arguments was that the targets were too small. The Hiroshima and Nagasaki bombs brought two cities—and a country; Japan surrendered four days after Nagasaki—to its knees. What would a hydrogen bomb have done? As horrible as they were, the atomic bombs still left a semblance of cities and their populations—something to build on. Hydrogen bombs would have left a wasteland. Once the Ulam–Teller idea was known, the hydrogen bomb, in Oppenheimer's memorable and unfortunate phrase, became 'technically sweet'. It could not be stopped. While I think that an argument can be made for the making and using of atomic bombs during the war, I do not see any argument for making the hydrogen bomb. As far as I am concerned, it was a terrible mistake.

As I mentioned before, my own brief encounter with nuclear weapons took place in the summer of 1957—over a half-century ago. I found that when I went to reconstruct the events I had a faulty memory. The original Trinity test was photographed in millisecond intervals by Harold Edgerton, the famed expert in high-speed photography. You can see the whole sequence on the web. The camera was going so fast that only after about two-thirds of a mile of film had been shot did the explosion happen. One frame per millisecond is too slow to record the actual fission, which takes place in a microsecond. At this point the fireball is a small incredibly hot blob, with a temperature of something like 60 to a 100 million degrees Celsius. By comparison, the surface temperature of the Sun is about 6,000 degrees (even the Sun's center, at fifteen million degrees, is 'cooler' than the fission fireball). 'Brighter than a thousand suns' is about right, although at 60 million degrees most of the radiation is in the x-ray region and not in the visible.

Up to this point in the time sequence, all the effects are intrinsic to the bomb, be it fission or fusion. They do not depend on the environment. The same effects would occur in the vacuum of outer space. But what happens in the next fractions of a millisecond does depend on the environment. I am going to focus on the kind of ground-level burst that I saw. These take place in the Earth's atmosphere, which is mostly nitrogen and oxygen. These gases absorb the x-rays from the blast within a few feet of the center of the explosion. This heats up the air, which now acquires a temperature of some 300,000 degrees. The temperature falls rapidly with the expansion of the fireball. But even at the reduced temperature the pressure inside the fireball is a few thousand times the ambient air pressure. The air near the surface of the fireball is sharply compressed and this produces the shock wave. It moves much faster than the speed of sound in air and heats up the air it comes in contact with, making it incandescent. This incandescent air absorbs the much hotter radiation coming from the interior of the fireball, which now becomes invisible to an outside observer. At this stage the photographs show the incandescent air external to the fireball. At about 15 ms for a bomb the size of the gadget, the temperature of the air drops to where it is no longer opaque to the radiation from the fireball and at this stage Edgerton's photographs show the interior of the fireball with all its complexity. This is what I saw when I first turned around to face the blast.

Chapter 6

Centrifuges and proliferation: India and Pakistan

To make a Hiroshima-type nuclear weapon requires about a 100 pounds of uranium-235 in metallic form. We want it to be metallic so that we can shape it. We do not, it turns out, need absolutely pure uranium-235. If we can enrich a sample of natural uranium so that it becomes something like 90% uranium-235, that will work. But how to do this?

The method of producing plutonium has not changed since the 1940s. Reactors then, reactors now. The method of enriching uranium has changed, however. Centrifuges, the method of choice now, were tried and abandoned in World War II because they were too complicated and unreliable at that time. Instead, most of the budget and a large part of the real estate of the Manhattan project was devoted to producing U-235 by a method called gaseous diffusion.

The people at Oak Ridge did not have available to them what is known as the 'Zippe centrifuge'. The story of the Zippe centrifuge is so bizarre that I doubt that you could make it up. At least I could not. First let me explain how a centrifuge works. It is basically a device with a cylindrical container that can be made to spin around an axis. When it is spinning a centrifugal force acts on the molecules in the container. This force acts outwards and tends to push the molecules towards the outer surface of the container. The magnitude of the force depends on three things. It is proportional to the distance of the molecule from the axis of spin. It is proportional to the square of the rate of spin—the number of revolutions per second. Finally, and crucially, it is proportional to the masses of whatever the mixture of spinning particles is made up of. Thus, if you have a lighter and heavier component—e.g., cream and milk or compounds of U-235 and U-238—the heavier component will have more force on it and will move out to the sides, leaving the lighter component close to the middle. When I was preparing this chapter I decided to look up 'centrifuge' online to see what was on offer. I came across one called the 'Valcon QM100 Quantum Large Capacity Benchtop Centrifuge'. I liked the name.

doi:10.1088/978-0-7503-1308-7ch6

Anything with 'quantum' in it cannot be all bad. It goes for 1,366 US dollars. The manufacturers claim it can rotate at 2,800 revolutions per minute. This sounds impressive until one learns that the Zippe centrifuge can rotate at 90,000 revolutions per minute. What is its origin, including its name?

When, in the spring of 1945, the Soviets occupied Berlin, they knew a good deal about the atomic bomb programs—both ours and the Germans'. Los Alamos spies like Harry Gold, Ted Hall and, above all, Klaus Fuchs had turned over a lot of material, including the fact that the 'gadget' was going to be tested and that it was a plutonium device. The Soviets also knew which German scientists were working on nuclear energy and where they could be found. They got hold of as many as they could and shipped them east, along with their equipment. Among them was the 1925 Nobel Prize winner in physics, Gustav Hertz. Less well known, but very important, was a physicist named Max Steenbeck, who had headed the research department in an aircraft company. He, and the rest, ended up in a sort of gilded cage in the resort town of Sochumi on the Black Sea in Georgia. The task that the Sochumi inmates were given was the separation of uranium isotopes. Steenbeck's group, some 60 strong, explored separation by centrifuge. He made great improvements in the centrifuges that were then available. In 1946, he was joined by an Austrian physicist named Gernot Zippe. Zippe had been in a prisoner of war camp from which he was fished out. Zippe then took over the experimental side of the centrifuge development while Steenbeck focused on the theory. It is quite unclear who did what when it came to the final product and there was some contention when, in 1956, Zippe left Russia with the plans and his name attached to the device. It was not until he attended a scientific meeting in Amsterdam the next year that he realized that his centrifuge design was better than anyone else's.

The Zippe centrifuge is what is known as a 'gas centrifuge', meaning that the material that is being separated is in gaseous form. A Zippe centrifuge is a thin tube, which, in the original version, was made out of aluminum. ('Aluminum tubes' sounds familiar from an absurd American intelligence mistake that was made to help justify the war in Iraq.) An improved version uses special steel, or most recently carbon fiber tubes. One of the innovations in the Zippe centrifuge is to introduce heating elements at the bottom of the cylinder, which drives the gas up and increases the efficiency of the separation. In the case of uranium, natural uranium is converted into a gas by combining it with fluorine. One uranium atom combines with six fluorine atoms to make a molecule of uranium hexafluoride. This is a difficult gas to make and a dangerous one to use. A country such as Iran that has manufactured sizeable amounts of uranium hexafluoride already demonstrates technological sophistication. As I have already mentioned, natural uranium is over 99 per cent uranium-238. Molecules of UF6 with U-235 and U-238 differ in mass by less than one per cent. Hence the separation is not easy. The centrifuges are configured in a cascade in which each partial separation is fed into the next step. A cascade of, say, 3,000 centrifuges, running for a year, will produce enough highly enriched uranium to make at least one bomb. But before talking more about bombs, let me tell you what happened to the Zippe centrifuge.

Upon his return to Germany, Zippe went to work for the Degussa Corporation, one of the largest chemical companies in Germany. Many German companies behaved disgracefully during the Nazi period, and Degussa was one of the worst. Their subsidiary, Auer, made, and sold, uranium metal to the German nuclear energy program. The most dangerous part of making this was done by slave laborers from concentration camps. One of its subsidiaries, Degesch, produced, and sold, the Zyklon B gas that was used in the German extermination camps. And, to complete the cycle, between 1939 and 1945 they received at least five metric tons of gold, much of it extracted from the teeth of concentration camp victims who had been murdered with Zyklon B. This was the company that Zippe joined. In 1958, Zippe went to the University of Virginia to work with Jesse Beams, who was the first person to separate isotopes with a gas centrifuge. In 1934 he had separated isotopes of chlorine. There was now a four-way cooperation between Degussa, the University of Virginia, our Atomic Energy Commission and the German Ministry for Atomic Energy. A German state-owned corporation was formed, which was privatized in 1970. In 1971, this company was merged with a British and a Dutch company. The merged company was known as URENCO. Their goal was to produce Zippe centrifuges for industry. It is the Dutch company that interests us. It had a subcontractor located in Almelo in Holland. In May 1972, it acquired a new employee, a young Pakistani metallurgist named Abdul Qadeer Khan.

Khan was born in 1936 in Bophal, which was then in British India. After partition he made a very traumatic move to Pakistan. He did his undergraduate work in Karachi and then went to West Germany to continue his studies, moving to Delft in Holland and, finally, Belgium, where he obtained his PhD. His thesis advisor got him his job in Almelo. Along the way, he had acquired Dutch and German language skills and a South African-born Dutch wife. In 1971, the Indians supported a successful rebellion in East Pakistan, which created Bangladesh. That year the western-educated Zulifkar Ali Bhutto became president of Pakistan and was able to implement his longstanding idea of creating a nuclear deterrent. When the Indians successfully tested an atomic bomb in 1974 he famously said that the Pakistanis would have a nuclear weapon even if it meant that they had to eat grass. From the pictures I have seen of Bhutto, I do not think he deprived himself of nourishment. He called for scientists in the diaspora to return and help. Khan wrote a letter directly to Bhutto offering his services, which were accepted. He decamped from Holland with his wife and the plans for the Zippe centrifuge. At first, he was content to work within the official program, but when he decided it was moving too slowly, he was able to persuade Bhutto to set him up in his own enterprise near Islamabad, which eventually became the Khan Research Laboratories. It was so closely held that even government officials were not allowed to enter it. It became the perfect cover for Khan's activities. I will just sketch these. He began by trading the centrifuge design to the Chinese for plans for a fission bomb—successfully tested in 1998—which the Chinese had obtained from the Soviets, who had in turn obtained early versions from Los Alamos via Fuchs. Khan traded centrifuges to North Korea for missiles. At the same time Degussa was trying to sell centrifuges to Iraq, he sold them to Iran, where they formed the basis for the Iranian enrichment program. He sold the whole package, centrifuge parts, bomb plans and all, to Gaddafi in Libya, but the ship bringing some of it from Malaysia in October 2003 was intercepted and Gaddafi turned everything over to the CIA. In February 2004, Khan confessed all of this to Pervez Musharraf, then president of Pakistan, and was put under house arrest in Rawalpindi, where he remains until the present day—thanks to the Zippe centrifuge.

* * * *

Homi Jenhangir Bhabha, the scientist largely responsible for India's atom bomb, was born in Bombay on 30 October 1909. He came from a Parsi family closely associated with the Tatas, one of the richest families in India. His father was a British-educated lawyer for the Tatas and his paternal aunt had married Sir Dorab Tata, the oldest son of the founder of the Tata industrial empire, Jamshetji N Tata.

Bhabha grew up in an atmosphere of high culture and great privilege. In his aunt's house could be found people like Nehru and Gandhi. He was educated in the best of the most expensive schools in Bombay. He showed great natural abilities in mathematics and he persuaded his uncle to finance studies at Cambridge in the UK. He was supposed to become an engineer, but at Cambridge he attended the lectures of P A M Dirac, one of the founders of quantum theory, and decided to take his degree in physics. He did his PhD with Ralph Fowler, who had been Dirac's mentor. As a postdoctoral scholar he studied with the best people in Europe. During this time he did a calculation of the collisions of electrons with positrons—a process that bears his name and can be found in textbooks. When he was on vacation in India in 1939 the war in Europe broke out and Bhabha decided that he could not go back to Britain. This was a momentous choice, which eventually led to the creation of the Indian atomic bomb.

Ali Bhutto, the politician who organized Pakistan's nuclear weapons program, was born on 5 January 1928 in Larkana, then part of British India but now in Pakistan. His father, Sir Shah Nawaz Bhutto, was a landowner in the Sindh. While not in the same financial league as the Tatas, the Bhuttos were still very well off. The family was Muslim and his father was an important political leader. Like Bhabha, Bhutto received his primary education in private schools in Bombay. Also like Bhabha, he left the Indian subcontinent to study abroad. He spent two years at the University of Southern California before transferring to Berkeley. He then went to Oxford in the UK to study law. He practiced law in Pakistan briefly before entering government. In 1957 he became the youngest member of the Pakistani delegation to the United Nations.

Six years later he became the Foreign Minister and angered the Americans by developing a close relationship with China. One of his friends was a nuclear engineer named Munir Ahmad Khan—no relation to A Q Khan—who had worked extensively in the United States but took a position as a technical associate with the International Atomic Energy Agency in Vienna. This gave Khan access to classified reports and in 1965 he informed Bhutto that India was starting a serious program to build an atomic bomb. Bhutto reacted by saying, 'Pakistan will fight for a thousand years. If India builds a bomb Pakistan will eat grass or leaves or even go hungry, but we will get one of our own. We have no other choice.' Munir Khan came back to Pakistan to lead the project that has by now constructed about 100 such weapons, matching India's supply. The two countries are like scorpions in a bottle. If one stings the other they will both die. A full-scale attack would reduce both countries to rubble.

While much has been written about the nuclear bombs of India and Pakistan, there is nothing like the collection of essays entitled *Confronting the Bomb,* by seven Indian and Pakistani scientists, with an introduction by the chemistry Nobelist John Polyani [1]. The collection was edited by the Pakistani physicist Pervez Hoodbhoy. Born in Karachi in 1950, he recalls that as a young boy he went to schools in Pakistan and was later tutored by the physicist Abdus Salam, who worked on the Pakistani bomb and is the only Pakistani to have won a Nobel Prize in any subject. Hoodbhoy took his degrees from MIT and had various teaching and research appointments until he joined the faculty at the Quaid-i-Azam University in Islamabad. Apart from physics, he is widely known for his courageous and outspoken struggle against Islamic extremism in Pakistan as well as his opposition to nuclear weapons.

The first essay in the book—'Scientists and India's nuclear bomb'—is by the Indian physicist M V Ramana, who is currently at Princeton University. Among other things, Ramana gives an historical account of the Indian nuclear program. Fission, as I have discussed, was discovered in Germany at the end of 1938 and its implications soon became clear. They were understood in India, and in 1948 the newly independent India passed an Atomic Energy Act and created a commission to supervise atomic energy. Bhabha was appointed chairman. In fact, Bhabha had already written to his uncle Dorab Tata's trust in March 1944 to ask for money to begin a nuclear energy program and fund the Tata Institute of Fundamental Research in Mumbai—then Bombay. This date is surprising, since it was over a year before the US set off its first nuclear explosions. What inspired Bhabha to make this request at that time? He did envision a nuclear power program and believed that India should have a homegrown group of experts to run it. Did he know the details of the US program to make nuclear weapons, perhaps through his British contacts? Once the bombs were dropped on Japan, Bhabha's intentions were clear.

I think of Bhabha as in some ways the Edward Teller of India. Teller was obsessed with the Soviet Union and Bhabha was obsessed with China. He thought that the only way to deter China was to have a nuclear arsenal and he was prepared to do anything to obtain one for India. He had support from Nehru, who tried to disguise the bomb program as a means of using nuclear power—including atomic explosions—for peaceful purposes. Teller also tried to persuade people that nuclear explosions were important for projects like dredging harbors.

Atoms for Peace

For India and Pakistan, nuclear reactors provided the path to nuclear weapons, just as Fermi's 'pile' had done in the United States. As I have noted, the average fission, whether in a reactor or a bomb, produces two free neutrons. Depending on the size, shape and density of the fuel, as thousands of fissions take place per second, some free neutrons will escape through the surface of the fuel and some will bounce around within it, potentially colliding with other fissile nuclei and causing them to split in turn. A chain reaction depends on making sure that not too many neutrons are allowed to escape and that there are enough fissile nuclei nearby to make such collisions likely. When the ratio of neutrons being produced within to those escaping through the surface of a mass of fuel is one to one, the mass is said to be critical. In this state, it is on the cusp of being able to sustain a chain reaction, but it will not do so for long. This is because the supply of fissile atoms in a critical mass of fuel is constantly being consumed, and as this happens it becomes ever more likely that neutrons will simply be absorbed by non-fissile isotopes in the fuel instead of contributing to the chain reaction. A truly self-sustaining reaction therefore requires a slightly 'super-critical' mass of fuel—where more neutrons are created than escape.

As we have seen, creating a super-critical mass is difficult because the isotope that accounts for 99.3 per cent of natural uranium, uranium-238, is not fissile—that is, it can only be split by a neutron traveling at relatively high speed, not by the fraction of slow-moving neutrons produced by fission. Ordinary uranium (without a moderator) cannot sustain a chain reaction, and in fact its U-238 nuclei tend to inhibit the reaction by absorbing free neutrons. Most contemporary reactors require fuel in which U-235 is enriched from its natural 0.7 per cent to at least three per cent, and weapons-grade uranium must be enriched to about 90 per cent. As I have noted, such concentrations can be achieved by using centrifuges to separate isotopes according to minute differences in mass, but this is a complex and expensive process, beyond the means of many countries and difficult to carry out in secret.

In December 1953, President Eisenhower announced the Atoms for Peace program, which was intended to share peaceful nuclear research and technology with America's allies and provided a large number of developing countries, including Iran and Pakistan, with their first reactors. One oddity of these reactors is that they were fueled by weapons-grade uranium, i.e., uranium that had a concentration of at least 90 per cent U-235, which was available at the time because a surplus of it was created by the US weapons program. No one thought that these countries would ever have the ability to make nuclear weapons—and the United States provided weapons-grade uranium to its Atoms for Peace clients until 1978. Many years earlier, President Truman had announced that the Soviets would never have the ability to make a nuclear weapon.

In August 1956, Indian scientists began operating a very small reactor at a new research facility in Trombay, known as the Atomic Energy Establishment. (It was renamed for Bhabha after his death in 1991.) This reactor used 80 per cent enriched uranium supplied by Britain. But a different path in India to a weapon—using plutonium instead of uranium as the fuel—was opened by a Canadian-designed reactor that was begun in 1954 and first went critical in 1960.

In a chain reaction, the neutrons emitted by fissioning nuclei must be slowed down by a moderator, such as water or graphite, in order to increase the probability of their fissioning the fissile isotope U-235. In the example I discussed earlier the moderator was graphite. In the case of the Canadian reactor, the moderator was heavy water, in which ordinary hydrogen, with a single proton, is replaced by a relatively rare isotope that also contains a neutron in its nucleus—that is, deuterons. Light water would be ideal, except that it absorbs neutrons in a process in which a proton captures a neutron and a deuteron is formed. Deuterons have a much smaller capture rate. Light water can be used if enriched uranium is used as the fuel. What is crucial to plutonium production is that with a heavy water moderator one can use fuel with the low U-235 concentration found in natural uranium. (This is what the Germans had hoped to do.)

During the chain reaction the U-238 that constitutes a significant portion of any reactor fuel is constantly being converted into U-239 by the absorption of neutrons. The nuclei of this new isotope are unstable and will rapidly emit electrons and anti-neutrinos—a process known as beta decay—to become neptunium-239, yet another unstable isotope that in turn undergoes beta decay, yielding plutonium-239, a fissile isotope that can fuel both reactors and weapons. Thus, by choosing a reactor that used natural uranium that is 99.3 per cent U-238 and a heavy water moderator Bhabha and his colleagues maximized the amount of fissile plutonium that could be extracted from the byproducts of their supposedly peaceful energy program.

The Canadians were suspicious of the Indians' intentions, but Bhabha refused to accept any kind of inspections. Reasoning that the Indians would get their reactors from somewhere, the Canadians decided that they need not lose the sale. Construction of the Indian reactor began in 1955 at Trombay. Some of the heavy water was supplied by the United States. Despite assurances that they would not pursue weapons, the Indians began extracting plutonium from the reactor using chemical procedures that they had also learned in the United States.

In 1962, there was a war between India and China, ostensibly over territory along their border in the Himalayas. Two years later, the Chinese tested their first bomb. After this, Bhabha claimed that India could produce a stockpile of 50 atomic weapons for as little as 21 million US dollars—about 165 million in today's dollars—a misleading extrapolation from American estimates of the cost of building a single device using proven designs and pre-existing facilities. (Conservative estimates of the cost of the Indian weapons program to date are in the tens of billions of dollars.) Under the guise of a peaceful nuclear energy program, India designed a bomb and accumulated enough plutonium to set off its first test device in 1974.

In an essay entitled 'Pakistan: climbing the nuclear ladder' [1], Hoodbhoy gives an account of the history of nuclear weapons in Pakistan. I was a visiting Ford Foundation professor at the University of Islamabad in the fall semester of 1959. I was taken to visit the first reactor to arrive in Pakistan. It was part of the Atoms for Peace program and had been built by the American Machine and Foundry Company. Abdus Salam went to the United States in 1965 as part of a small delegation that persuaded Edward Durrell Stone to design the campus of the Pakistan Institute of Nuclear Science and Technology. It looked more like a religious shrine than a nuclear facility.

The first Pakistani reactor had a small power output—about five megawatts as compared with the thousands of megawatts that an electricity-generating reactor produces. As with a number of other Atoms for Peace reactors, its fuel was weapons-grade uranium, enriched to 93 per cent. Due to the low proportion of U-238 in its fuel, this reactor was not optimal for making plutonium and was used to create isotopes for medical use and to train physicists. In 1991 the reactor was redesigned to operate with 20 per cent enriched uranium, and it seems that some plutonium production now takes place on the site using the successors to this reactor.

Atoms for war

During my tenure in Islamabad, my host was one of Hoodbhoy's predecessors, Riazzuddin, who used only one name.

His twin brother Fayyazuddin was also a professor. They struck me as wonderfully kind and unworldly. Invited to dinner at Riazzuddin's, I arrived to find myself and Fayyazuddin the only guests. Apart from serving the food, the women in the household disappeared. After the meal I was expected to leave, which I did.

I was extremely surprised to learn later that Riazzuddin subsequently headed the team of theoretical physicists that designed the first Pakistani atomic bomb. Fayazzuddin refused to work on the bomb. In December 1973 Riazzuddin accepted a fellowship at the University of Maryland near Washington, DC and spent much time in the Library of Congress collecting all the unclassified material he could find on the design of a nuclear weapon. As I discussed earlier, to make a nuclear explosion you must very rapidly create a highly super-critical mass in which the rate

Figure 6.1. Riazuddin. Credit: Wikimedia Commons. This image has been obtained by the author(s) from the Wikimedia website https://commons.wikimedia.org/wiki/File:Riazuddin.jpg, where it is stated to have been released into the public domain. It is included within this article on that basis.

of fission increases exponentially. We have seen that one way of doing this is to implode—squeeze down—a sphere of plutonium or uranium, thus increasing the density of the sphere, which lowers the critical mass. Riazzudin found unclassified information about how to do this—including reports from Los Alamos on Robert Oppenheimer's design for the first implosion device. Thanks to the machinations of the Pakistani physicist A Q Khan—no relation to Munir—the group also obtained some details of the Chinese design. They had already done the basic work on their implosion design, but the American and Chinese information helped them improve it. The Russians had not yet released the material that Klaus Fuchs had given them during the war.

A Q Khan also worked in the Netherlands from 1972 until 1975 for the Urenco Group, a commercial supplier of enriched uranium, where he stole the plans for the advanced centrifuges used to produce the explosive fuel that could be used for Pakistani nuclear weapons. Khan had direct contact with Bhutto, who gave top priority to the creation of this facility. (The Urenco design was later given to North Korea by the Pakistanis in exchange for missile technology, and centrifuges based on it were found by IAEA inspectors in Libya and Iran.)

If Bhabha was the Edward Teller of the Indian program, one can make the case that Abdus Salam was his Pakistani counterpart. Having had personal dealings with both Salam and Teller, I have no doubt that Salam was the more agreeable of the two. He was born in 1926 in the Punjab in British India. His family was of the Almadiyya sect, which was later declared to be non-Muslim in the Pakistani constitution. He was a brilliant student and was headed for a career in the Indian Civil Service—the part devoted to railways. But for some reason this involved an eye examination, which Salam failed. He went on to study mathematics in the graduate school in Lahore and then won a scholarship to Cambridge in the UK. His PhD thesis involved problems concerning infinities in quantum electrodynamics, which Freeman Dyson had left for others to solve.

He returned to Pakistan to teach, but in 1957 he went to Imperial College in London. Salam was a great teacher and collaborator. He worked with Steven Weinberg on what is known as the Higgs mechanism for generating mass. He also did independent work on neutrinos. He shared the 1979 Nobel Prize with Weinberg and Sheldon Glashow. Both Fayyazuddin and Riazzuddin were his students and Hoodbhoy also worked with him. In 1964 Salam founded the International Centre for Theoretical Physics in Trieste, which welcomed young physicists from developing countries. In 1972 he was called back to Pakistan by Bhutto to begin work on the bomb. I do not know if Salam actually took part in the technical work that led to the bomb, but he did troll for unclassified literature and above all he recruited his students to work on the project. His old high school friend Munir Khan was appointed chairman of the Pakistan Atomic Energy Commission. I have heard that the money Bhutto raised to build a bomb came from Saudi Arabia and Libya, apparently on Pakistan International Airlines flights in bales filled with dollars.

On 28 May 1998 the Pakistanis exploded five atomic bombs underground. Prime Minister Nawaz Sharif declared 'Today we have settled scores with India by detonating five nuclear devices of our own. We have paid them back.' By then both Salam and Bhutto had died. Salam's remains were brought back to Pakistan from Oxford. He was buried next to his parents in an Ahmadiyya cemetery. His gravestone has now been defaced.

For many people, including myself, the most urgent question is the security of the Pakistan nuclear arsenal. This is addressed in another essay by Hoodbhoy entitled 'Post bin Laden: the safety and security of Pakistan's nuclear arsenal' [1]. Whatever one may think of General Pervez Musharraf, he made a serious attempt to take control of the Pakistani nuclear arsenal. In February 2000 he created something called the Strategic Plans Division, which has some 12,000 personnel and is in charge of the nuclear weapons. The problem with the security of all such entities in Pakistan is the presence of extremists.

In his post-Bin Laden essay Hoodbhoy gives an example. In 2003 Musharraf escaped two assassination attempts by military officers with extreme Muslim views. They were caught and some were sentenced to death. In May 2012 they all escaped from jail while the prison guards stood aside and displayed Taliban slogans. In Hoodbhoy's view, there are in effect two Pakistani armies. One is devoted to defending the country's borders and another is devoted to turning Pakistan into a country ruled by Shari'a law. The great concern is that this army will get hold of nuclear weapons. A relevant question is whether Pakistan has a system requiring that two or three separate officials give assent before a nuclear device can be launched. If this had been in effect in *Doctor Strangelove*, General Ripper would not have been able to launch his strike. Who in Pakistan has the authority to launch? Who in India or Israel for that matter? That is of course part of the concern about Iran. Who, besides the Supreme Leader, could launch a bomb if Iran ever gets one? From all the reports I have seen, no one outside the system knows where the Pakistani arsenal is maintained. Some of it, Hoodbhoy thinks, may be on unmarked trucks that move around within and between cities. How can one be sure that all the drivers and guards can be trusted? Extremists respect martyrs. Could an entire city be cast as a martyr?

Reading Hoodbhoy's book does not make for optimism about the future. Both India and Pakistan are building nuclear weapons at a seemingly increasing rate. No one sees where it will end.

Reference

[1] Hoodbhoy P (ed) 2013 *Confronting the Bomb* (Karachi: Oxford University Press)

IOP Publishing

One Physicist's Guide to Nuclear Weapons
A global perspective
Jeremy Bernstein

Chapter 7

Israel, North Korea and Iran

In most of the Western world, it is considered okay for Israel to have nuclear weapons, not okay for North Korea to have them, and not okay for Iran to acquire them. In this chapter I discuss briefly the nuclear situation in these three very different countries. In chapter 9 I continue the discussion of Iran and also have a brief look at the nuclear prospects in two other countries and in terrorists groups ('non-state actors').

In the foregoing chapters I described attempts to make nuclear weapons both successful (US and UK, India and Pakistan) and not (Nazi Germany). What all these attempts have in common is that very considerable resources were devoted to them. The same is true of the Soviet program, which I have so far mentioned only briefly. Once Stalin decided that the Soviets had to have a bomb, all the necessary resources were commandeered. Keep in mind that the Soviet program started at the end of World War II, when much of the country had been devastated. Nonetheless the USSR created the needed industry, including a plutonium-producing reactor, with such speed that a successful test was conducted some four years after Hiroshima. While the information they received from Fuchs and others was useful, it was not necessary. They would have obtained their fission bomb on their own, maybe a little later, and it probably would have had a better design. The Soviet physicists knew the weaknesses of the design that they had received from Fuchs, but were ordered to duplicate it. The original bombs were very inefficient. With redesigned pits, efficiency could be greatly improved. The Soviets did develop the hydrogen bomb on their own, also with massive deployment of resources. They learned little of significance from espionage. What about other countries?

doi:10.1088/978-0-7503-1308-7ch7

Israel

In 1960 I made a trip to Israel. I had been working in Paris with Murray Gell-Mann just prior to the time when he discovered the quark. Our trips to Israel were totally independent and I was as surprised to see him there as he was to see me. We joined forces for a few outings. I had a letter of introduction to Renana—the daughter of Ben-Gurion. This led to an invitation to visit Ben-Gurion at his desert retreat in the Negev S'deh Boker. A South African physicist I had known earlier, and who had emigrated to Israel, drove us. I think Ben-Gurion had been briefed about Gell-Mann as a future Nobelist, because he delivered a pitch for emigration. Among other things, he pointed to a hill and asked Gell-Mann, 'Do you see those trees?' There were no trees and Ben-Gurion said, 'You come back and in ten years, there will be trees.'

On the way back to Tel Aviv at night we got lost. At one point I looked out the window and saw barbed wire and lights. Then I saw a road sign that said 'Dimona'. We beat a hasty retreat. If we had gotten in we would have found that the dominant language was French. The French were playing an essential role in constructing the plutonium-producing reactor at Dimona that is the basis of the Israeli bomb program. It is true that in 1960 de Gaulle attempted to shut down the cooperation, but it continued until 1966. The buildings of the project were brightly illuminated and close by.

The second recollection is of an event that occurred in New York. The Israelis have said again and again that they will not be the first to 'introduce' nuclear weapons into the Middle East. What is meant by 'introduce'? This anecdote may help to adumbrate. From 1984 to 1988 Benjamin Netanyahu was the Israeli ambassador to the United Nations. During this time I was invited to a small party where he was one of the guests. The hostess introduced me as a nuclear physicist who had been to Los Alamos. It is true that I had gone to Los Alamos for a few months in the summer of 1957 and that I had written a couple of papers that involved nuclei, but that was about it. Yet it was enough to capture Netanyahu's attention. I took the opportunity to ask him about the Israeli nuclear weapons program. He said they had them and would use them *in extremis*. I am sure nothing has changed except that they now have more of them and improved methods to deliver them.

North Korea

The major North Korean nuclear facility—the Yongbyon Nuclear Scientific Research Center—is located about 40 miles north of the capital, Pyongyang. On the site there are the nuclear reactors that produce plutonium and a reprocessing facility for recovering the plutonium from the reactor fuel elements.

There is also a centrifuge facility for enriching uranium that operates with centrifuges that are more advanced than the ones in general use in Iran. Some of the scientists who operate the Yongbyon facility were originally trained in the Soviet Union, but the present generation appears to have been trained in the country. The construction of the first reactor began in 1980. The choice of its fuel was dictated by the limitations of the country. At the time there was no enriched uranium and no supply of heavy water. The North Koreans copied the design of the first British plutonium-producing reactors, which was available as an open source. These are called 'magnox reactors'. Magnox is an alloy of magnesium with a small amount of aluminum. In these reactors it is used to clad the fuel elements, which are made of natural unenriched uranium. All reactors need a 'moderator' to slow the neutrons produced in fission. In these reactors graphite is used. These reactors were long ago abandoned in Britain, but they continue to function in North Korea. It should be noted that reactors that use unenriched uranium for fuel are ideal for manufacturing plutonium. That is because this uranium is over 99 per cent uranium-238 and the plutonium-producing process begins when a uranium-238 nucleus absorbs neutron. The Arak reactor in Iran, which was to be moderated by heavy water, was also designed to use natural uranium fuel. Unlike in reactors that are used primarily to generate electricity, the fuel elements of these reactors are designed so that they can be changed readily—every few months—so that the plutonium can be extracted. Leaving them in too long produces unwanted isotopes that make the product less suitable for weapons.

The existence of the centrifuge facility was revealed in November 2010. The American physicist Siegfried Hecker and his two colleagues John Lewis and Robert Carlin were on an official visit. Hecker had been for many years the director of the Los Alamos laboratory. Hecker described the experience: 'I was stunned by the sight of 2,000 centrifuges in two cascade halls and an ultramodern control room [...] Although I and other nonproliferation experts had long believed that North Korea possessed a parallel uranium-enrichment program—and there was ample evidence for such a belief—I was amazed by its scale and sophistication. Instead of finding a few dozen first-generation centrifuges, we saw rows of advanced centrifuges, apparently fully operational. Our hosts told us that construction of the centrifuge facility began in April 2009 and was completed a few days before our arrival. That is not credible, however, given the requirements for specialty materials, as well as the difficulty of making centrifuge cascades work smoothly.'

Exactly how North Korea gathered both the materials needed and the plans to build such a facility is not clear. It is known that beginning in the early 1990s with the collusion of Pakistan's government the Pakistani proliferator A Q Khan exchanged centrifuge technology for North Korean missiles. The exchange was facilitated by the use of Pakistani military aircraft for transport. In around 2000 some 24 Pakistani centrifuges were delivered to North Korea. These were presumably of the old type and it does not explain how by 2010 an ultramodern facility with thousands of centrifuges was in operation. Constructing such a centrifuge requires highly specialized materials, such as special steels. Where did these come from? Iran has been suggested. If this was the case what was the *quid pro quo*?

As of this writing, Hecker estimates that the North Koreans may have enough fissile material for 18 bombs with the capacity to produce six to seven a year. So far they have tested four devices, three of them certainly using plutonium. To put the yields of these tests in perspective, recall that the Hiroshima bomb was equivalent to about 15 kilotons of TNT. The North Koreans' first test on 9 October 2006 produced a one-kiloton explosion. If this was a failure—a 'fizzle' to employ the term of art—it is in the eye of the beholder. The second test was on 25 May 2009 and produced a yield of four kilotons, while the third on 12 February 2013 produced a yield of seven kilotons. Clearly their technology has been steadily improving.

The most recent North Korean test (as of this writing) was on 6 January 2016. It appears to have produced a yield of about ten kilotons. As I write this, it has not been announced whether this was a plutonium or a uranium device. What has been announced by Kim Jon-Un is that this was some sort of 'hydrogen bomb'. Let us recall what this means. The devices that were tested previously used fission as their energy source. But a hydrogen bomb uses the fusion of light elements such as the isotopes of hydrogen as its energy source, or at least one of them. Fission comes into play in two ways. A true hydrogen bomb uses a fission device as its trigger. This produces the temperatures and pressures needed to induce fusion, which produces very energetic neutrons that can induce more fission. The resulting yield can be in the megaton range. Therefore it is extremely unlikely that the North Korean device was a true hydrogen bomb. It was more likely what is known as a 'boosted device'. In such a device a fission explosion ignites a small quantity of thermonuclear fuel, which produces very energetic neutrons that cause more fission. This enhances the ratios of the yield to weight and volume of the device. The bombs can be made lighter. Hecker has stated that we may never learn exactly what the North Koreans tested. He noted in an interview, 'North Korea has now been in the nuclear testing business for almost ten years, so we can't rule anything out for certain.'

David Albright, who served as an inspector for the International Atomic Energy Agency in both Afghanistan and Iraq and who has studied nuclear proliferation closely has noted some characteristics of this 2016 test as opposed to the most recent previous one. This test occurred about 700–800 meters below a mountain as opposed to the 350 meters of the previous test. This is as if the North Koreans were expecting a high yield. The way the nature of a test is usually discovered after the fact is when the isotopes produced in the explosion seep into the atmosphere, where they can be detected. The delay in finding such markers until long after the previous test suggests that the North Koreans have taken pains to reduce this leakage. As of this writing, nothing has turned up from the January 2016 test. If this test had a fusion component it would mean that the North Koreans have been able to produce both deuterium (with a nucleus of one neutron and one proton) and tritium (with a nucleus of two neutrons and a proton). The most energetic fusion reaction involves the fusion of a triton and a deuteron to produce helium and an energetic neutron. All samples of natural water contain fractional amounts of heavy water. This can be separated out by electrical means. While the deuteron is stable, the triton is not; it decays with a half-life of a little over 12 years, and about 5.5 per cent of any sample is lost each year. Therefore it is not found naturally, but has to be manufactured. This can be done in reactors. One method is to produce lithium 6, which produces tritons when it is irradiated with neutrons. Albright has found evidence that the North Koreans have developed facilities for doing this. In short it would appear as if the North Koreans are determined to produce fuel to be used in fusion-enhanced nuclear weapons. This is a very serious matter because these weapons can be made light enough to fit on rockets, which the North Koreans have in abundance. This is the real threat.

Iran

To discuss Iran in any detail would require a full book [1]. This will be an abbreviated account (with some extended discussion in chapter 9). The Iranians have mastered the nuclear fuel cycle. They have taken uranium from mines and then converted it into uranium hexafluoride. This is the gas that is fed into centrifuges, of which the Iranians had some 18,000. With the nuclear agreement the number has been reduced to about 5,000. This gas is spun in cascades of centrifuges and the common heavy isotope U-238 is separated mechanically from the rare lighter isotope U-235—a process that is given the somewhat odd name 'enrichment'. The cascades can be set to enrich to any level desired. Above 20% is forbidden by treaty. The Iranians have produced enough 20% enriched uranium to provide the fuel elements for the Teheran Research Reactor, thus completing the cycle. It is clear that if these cascades were arranged to produce weapons-grade uranium they could produce enough for a nuclear device in a matter of months. It is also well known that the Iranians have tried to buy technology suited to weapons on the black market. What success they had I am not sure. It is also known that they began construction of a heavy water reactor at Arak which was ideal for producing plutonium. This has been dismantled because of the agreement and its replacement will use imported fuel, probably from Russia. It is also suspected that on the military base Parchin near Teheran implosion experiments were performed. The International Atomic Energy Agency was never allowed to make a proper inspection of this base and only limited information can be obtained from satellites. It would not be surprising if there were other facilities in Iran that could be devoted to this work. A massive complex is not necessary. At Los Alamos implosion tests were conducted in canyons not far from the residential area. Then there is the question of design.

In 1997 A Q Khan sold a rather advanced Chinese design of a uranium implosion device to the Libyans. We know exactly what this was because in 2003 they gave it up. I am sure the classified details are in some vault somewhere. But it was the design of the device the Chinese tested on 27 October 1966. It was flown on a rocket that could carry a payload of about 1,300 pounds. The bomb that was dropped on Nagasaki weighed over 10,000 pounds, so we are dealing with an advanced technology. I cannot imagine that A Q Khan did not sell this to the Iranians. He was always interested in making money and not much else. He will not say and the Iranians will not say, so I at least do not know for sure.

The joint nuclear deal attempts to confront all the aspects of the Iranian program. I for one was surprised by its detail and by its rapid implementation by the Iranians. The 18,000 gas centrifuges in Iran suitable for enriching uranium were located in two facilities, Natanz and Fordow, with the majority in Natanz. Fordow is located deep underground in a mountain near the sacred city of Quom. It has, in addition to centrifuges of the early type obtained from Pakistan, some more modern centrifuges whose enrichment capacity is much greater. The centrifuges are arranged in 'cascades' of 164. In such a cascade the outputs of the centrifuges are fed one into another for further enrichment. The deal allows only six cascades to remain in Fordow, two of which will operate without uranium while the other four remain idle. The facility will be turned into an international laboratory of some sort. In Natanz 30 cascades will be allowed, all of them of the non-advanced variety. Thus the total number of centrifuges has been drastically reduced. None of these will be of the newer type. They cannot be deployed. Of equal importance is what has happened to the reactor under construction in Arak. This reactor was very suspicious because it was designed to use a heavy water moderator and natural uranium fuel. Thus it was designed optimally for producing plutonium. Under the agreement the present core of the reactor has been destroyed and replaced by a design that will use ordinary water and lightly enriched fuel. This fuel will be supplied from outside the country and removed along with its plutonium content. This arrangement has worked very well for the Bushehr power reactor, where the fuel is being supplied by the Russians. What the Iranians have been intransigent about is revealing past military applications. We know that at various times there were attempts to design nuclear weapons, but we do not know how far these got. I think we are unlikely to learn. A very important part of the deal is the agreement of the Iranians to give up most of their lightly enriched uranium. They have about 15,000 kg and when the deal is fully implemented they will be left with only 300. This can be augmented slowly with the remaining centrifuges, so this must be watched carefully. The Iranians agreed to give up their 20 per cent enriched uranium, except for what is to be used in the Teheran Research Reactor, a small reactor located in a university in Teheran. Most of the elements of the deal sunset after a decade or so. But a decade in the Middle East is an eternity. In any event, the deal will stall the Iranian program, at least for a while.

There is another story to tell about Iran that begins in China. In 1951 the Chinese signed a secret agreement with the Soviets in which they traded uranium ore for assistance on their nascent nuclear program. In 1957 there was a second agreement in which the Soviets were to supply a sample bomb, which they never did. But meanwhile the Soviets had supplied advisors to help the Chinese with bomb technology. These were withdrawn, beginning in 1960. But by this time the Chinese were well underway. They had to choose between enriched uranium and plutonium for the fissile explosive and they decided on uranium. This meant that their program was not visible to us. Plutonium is made in large reactors and these are more-or-less impossible to conceal. They use a good deal of water, usually supplied by a nearby river. They emit steam via a smokestack. On the other hand, a centrifuge plant is not difficult to conceal. It uses rather little power and can be hidden underground as the two plants in Iran—Natanz and Fodrow—were. Thus when the first test of the Chinese uranium device at Lop Nor took place on 16 October 1964 we were caught off guard. I do not know the details of this device, except that it was a uranium implosion bomb. The same theories of implosion apply here as to plutonium. However, from what we do know of the fourth Chinese test which took place on 27 October 1966, we can guess that the 1964 device was very probably not a clone of the Christy gadget or of Joe 1. The fourth test took place on a rocket that could carry a payload of about 1,300 pounds, whereas the Nagasaki bomb weighed over 10,000. Clearly it was a new technology that enabled the downsizing of these bombs.

As I mentioned, the first idea at Los Alamos during the war was to use a hollow pit. For expedience it was decided instead to build the solid pit Christy gadget. After the war the hollow pit was revived and as few as two—as opposed to 32—detonation points sufficed. The tests I saw in Nevada in the summer of 1957 involved devices with hollow pits. Even in the Christy gadget the pit was wrapped in a heavy material called a 'tamper' which reflected neutron and enhanced the chain reaction, greatly reducing the critical mass. The critical mass of tampered plutonium-239 is about ten pounds, half that of an untampered sphere.

Whether the Chinese used this design in their 1966 test I do not know. What I do know is that in 1982 the Pakistani proliferator A Q Khan acquired the design of the 1966 device in a swap of nuclear technology with the Chinese. I also know that beginning in 1997 Khan sold this technology to the Libyans and in 2003 they gave it up. So we have in our possession somewhere this presumably classified set of plans for a Chinese nuclear weapon. We also know that in 1987 Khan sold nuclear technology to the Iranians for some ten billion US dollars. This became the basis of their centrifuge program. It is difficult to believe that they did not buy the same bomb designs. Neither they nor Khan will tell us.

How serious is this? It is not serious so long as Iran does not have the industrial capacity to build a bomb, including the fissile material. We know that they tried to buy exploding wire detonators on the absurd pretense that they would be used for mining. We also suspect that implosion experiments have been performed at the Parchin military base near Teheran. I would be very surprised if the Iranians did not have a group of theoretical physicists somewhere—perhaps Esfahan—working on design. Their physicists are off limits to interrogation. The terms of the Joint Comprehensive Plan of Action have gone into effect. The IAEA has announced that the Iranians have complied with the request to turn over information on their past activities. Not much was turned over and the real history of the Iranian program will probably never be written.

Reference

[1] See, e.g., my book: ed Bernstein J 2014 *Nuclear Iran* (Cambridge, MA: Harvard University Press)

Chapter 8

Scorpions in a bottle

The targets are too small
J Robert Oppenheimer

...but they cannot decide the outcome of a war, since atomic bombs are quite insufficient for that. Of course monopoly ownership of the secret of the atomic bomb creates a threat, but against it there exist at least two means: (a) monopoly ownership of the atomic bomb cannot last for long and (b) the use of the atomic bomb will be prohibited.

Joseph Stalin 1946

The USSR and Klaus Fuchs

After the atomic bombing of Hiroshima and Nagasaki in August 1945, Robert Oppenheimer became the avatar of the new nuclear age. He was perfect for the part. He had the somewhat anguished look of a sadhu and had even studied Sanskrit. The things he said seemed profound, even though one was not quite sure of their meaning. He enjoyed the role. He had quelled a rebellion at Los Alamos when after the German surrender some of the physicists wanted to stop making the bomb. He felt one had to see it through, if for no other reason than to see if it worked. President Truman decided he wanted to meet Oppenheimer. The meeting took place at 10.30 am on 24 October 1945. Truman wanted Oppenheimer's help in getting the May–Johnson bill passed; it would have given the military sole jurisdiction over atomic energy. It became clear that Oppenheimer was, to put it mildly, not enthusiastic. Then Truman asked Oppenheimer when he thought the Soviets would get the bomb. Oppenheimer said he did not know. Truman responded by saying that he did, 'never'. He became livid and threw Oppenheimer out of his office, saying afterwards that he never wanted to see 'that son of a bitch' in his office again.

doi:10.1088/978-0-7503-1308-7ch8

Truman was the one who had ordered the bombing of Hiroshima and Nagasaki. I believe, and will explain, that this exchange had profound consequences for the future of nuclear weapons. Oppenheimer reputedly said, 'Mr President, I have blood on my hands,' leading to the reaction of Truman.

The people associated with the creation of the bomb speculated as to when the Soviets would have one. General Groves, who headed the project, said 20 years. Some people thought two or three. The consensus among the physicists was for about five. The intelligence estimate was for eight. There was no question among them that the Soviets would build one. One of the reasons was that the American physicists appreciated the level of Soviet physics. For example, when it became clear in the 1930s to Victor Weisskopf (who later became second-in-command of the Theory Division at Los Alamos) that as a Jew he was going to have to leave Europe, he considered an offer from the University of Kiev—he had spent a fair amount of time in the Soviet Union—but decided against it and went instead to the University of Rochester in New York. He knew a great many of the Soviet physicists. The brilliant Soviet theoretical physicist Lev Landau—later a Nobel Prize winner—came and went from Niels Bohr's institute in Copenhagen. An interesting case is that of Pytor Kapitsa—another Nobel Prize winner. He received his education in Russia but then came to Cambridge in the UK, where he worked for a decade, even directing a laboratory there. In 1934 he returned to the Soviet Union to visit his parents and was not allowed to leave. The people at Los Alamos probably speculated—correctly—that he would be involved in any attempt to build a bomb. Also, the Soviets had published important work in fission.

As far as we know, there were three Soviet spies at Los Alamos, all unknown to each other as spies. David Greenglass was a machinist who worked on shaping the high explosives used to implode the plutonium. He delivered somewhat crude drawings of these explosive 'lenses' to the Soviets. Ted Hall was 19 when as a fresh Harvard graduate he came to Los Alamos in 1944. He supplied some material to the Soviets, but his main importance was to confirm some of the material that the more significant Soviet spy Klaus Fuchs had delivered. The Soviets were suspicious that what they were being delivered was a plant designed to throw them off. Confirmations like those of Hall were important. Even more important was the judgment of the very few Soviet physicists who were allowed to see the material that Fuchs supplied. They knew that what they were seeing was priceless.

Klaus Fuchs was born in Rüsselsheim, Germany in 1911. His father was a Lutheran pastor who taught at the University of Leipzig, which Fuchs attended. Fuchs became involved in politics and joined the student branch of the Social Democratic Party of Germany. Fuchs came to side with the communist wing of the party and was expelled. He joined the Communist Party in 1932 and was then was forced to leave Germany. He took exile in Britain and obtained his PhD in physics from the University of Bristol. He then did postdoctoral work in Edinburgh with the great German physicist Max Born, also a German refugee. In 1940 Fuchs was classified as an 'enemy alien' and sent to Canada. Rudolf Peierls, who with his fellow refugee Otto Frisch wrote the memoranda that started the British atomic bomb project, arranged for Fuchs to return to Britain and join him in the project. Peierls was then at the University of Birmingham and he and his wife Genia often took in physicists as boarders. Fuchs came to live with them. Genia complained that he never said anything and compared him to a music player that only produced a sound when it was fed a coin.

The British had decided to go the way of separating uranium isotopes and Fuchs and Peierls wrote a seminal paper on the subject. They concluded that the then available centrifuge technology was too primitive and focused instead on the diffusion of uranium hexafluoride gas. The gas is forced through tiny pores and the light isotope diffuses a little faster, allowing the separation. By 1942 Fuchs had made arrangements to transmit information to the Soviet Union. His plans were interrupted when Peierls was transferred, along with Fuchs, to Columbia University in New York to continue with this work. The odd thing is that Fuchs did not want to go from Columbia University to Los Alamos. He wanted to go back to Britain. A remarkable letter, which I will now quote, exists from James Chadwick, the discoverer of the neutron and the head of the British delegation. It was written to Peierls. There are some names of British physicists we do not need to identify. 'Bethe' is Hans Bethe, who was the leader of the Theory Division at Los Alamos, referred to here by its code name 'Y'.

July 14, 1944

Dear Peierls,

I have now had talks with both Kearton and Fuchs about the future of the New York section and in particular about their own positions. As a result, Kearton will approach Keith and Benedict with the object of getting a letter by one or both of them to Groves to say that the services of Fuchs and Skyrme are no longer required. It is possible that this matter was raised by Groves on a visit to New York earlier in the week, but I have had no news from him so far.

The position of Skyrme is quite clear. Bethe or Oppenheimer should write to Groves asking for his service in Y. Groves has provisionally agreed and there should be little delay over his transfer.

Fuchs' future is not so clear. I gave you the gist of a cable from Akers in my letter of July 11. I did not agree with the suggestion made in this cable that Fuchs was not required in England, but I wished to discuss the question with Kearton before I made up my mind. Kearton was very strongly of the opinion that Fuchs was quite necessary in England if work on any kind of diffusion plant is to continue.

I have now had a talk with Fuchs himself. He feels that he has a special contribution to make in England, whereas in Y he would be one of a number and can make no really significant difference to the work.

I agree completely with these views of Kearton and Fuchs, and I feel sure you also agree at least in principle.

I come now to the point of this letter. It would put me in a very awkward position if a request for Fuchs' services in Y were to be sent to Groves. If Groves were to agree I also should have to consent, for the consequences of refusing, on the grounds that he was needed in England for work which can have no significance for the war, might be quite serious. It would certainly cause great resentment in some quarters and our relations with the US on this project would be impaired. I should attempt to justify his return as being useful for the New York project, for after his experience here he could interpret their requests and help to direct UK work into directions of immediate interest to them. This argument would of course not be valid if a low-separation diffusion plant were to be started in England.

I therefore do not want Bethe to ask for Fuchs. Further than that, I want Bethe to say that Fuchs would not be especially useful in Y, if Groves asks if they want him, as he may. This means some tactful work on your part and I hope you will be able to do what is necessary by suggestion rather than direct action.

I have prepared the ground here and I think the matter can be arranged. I have stated that Fuchs could be useful in Y but that his special qualifications are not on the nuclear side but on the diffusion plant.

I know something of what is happening in London I want to keep the New York position as fluid as possible.

Yours sincerely,
J Chadwick

Somehow the matter was resolved and Fuchs joined the Theoretical Division at Los Alamos in August 1944. On a visit to his sister in Boston in February 1945 Fuchs made his first transmission regarding Los Alamos data to his courier Harry Gold. What has always puzzled me is in what form the information was taken out of Los Alamos. This transmission was relatively brief, but later there would be multiple pages. He certainly turned over these transmissions to Gold in written form, since Gold would have no understanding of them. I have been told that the security at the gate in Los Alamos was somewhat erratic. In fact, people left the laboratory with secret documents that they had forgotten to leave behind. But I doubt that Fuchs would have taken this risk. If he had been caught he would have been finished. He had a photographic memory and I think that it is more than likely that he put what he remembered into written form after he left the laboratory. In this case, he informed the Soviets about the discovery of spontaneous fission in plutonium and the switch to implosion. Hall had previously told the Soviets about implosion in less detail.

Until the use of the bombs on Hiroshima and Nagasaki the Soviet program had been somewhat casual. But now Stalin ordered a crash program to build the bomb. This was preceded by a detailed report by Fuchs on the plutonium implosion device, which he transmitted in June 1945. The Russians have now released this and I give a sample below.

Active material

The element plutonium of delta-phase with specific gravity 15.8 is the active material of the atomic bomb. It is made in the shape of a spherical shell

> consisting of two halves, which just like the outer spherule of the initiator, are compressed in a nickel-carbonyl atmosphere. The outer diameter of the ball is 80–90 mm. The weight of the active material including the initiator is 7.3–10.0 kg. Between the hemispheres is a gasket of corrugated gold of thickness 0.1 mm, which protects against penetration of the initiator by high-speed jets moving along the junction plane of the hemispheres of active material. These jets can prematurely activate the initiator.
>
> In one of the hemispheres, there is an opening of diameter 25 mm, which is used to insert the initiator into the centre of the active material, where it is mounted on a special bracket. After inserting the initiator, the opening is closed with a plug, made also of plutonium.

One of the most significant revelations in this document has to do with the delta phase of plutonium. After Glenn Seaborg and collaborators first observed plutonium made in a cyclotron in 1941, micrograms became available for experimentation. William Zachariasen, a crystallographer working at the University of Chicago, found that the element existed in several 'allotropes'. These are different physical forms, such as diamond and graphite for carbon. The alpha phase is stable at room temperature, but is a chalk and useless for making a bomb. The delta phase is metallic but unstable at room temperature. It had to be alloyed with gallium for it to work. Being told that the plutonium used to make a bomb was in the delta phase saved the Soviets a great deal of time. As did the rest of what Fuchs told them. Once they decided to make a bomb they created an organization to do it. It was headed by Lavrenti Beria, Stalin's much-feared chief of the secret police. The scientific director was the physicist Iulii Borisovich Khariton. He had access to all of the espionage data, which he decided to confirm independently. He was ordered by Stalin to produce a clone of the plutonium bomb that flattened Nagasaki. Kapitsa contacted Stalin to suggest alternative ways of doing the job. He was summarily thrown off the project and placed under house arrest, where he remained until Stalin's death. The physics was done in a new facility that had been built in the Urals town of Arzmas, which the scientists came to call Los Arzmas. Since they had decided to go the plutonium route they had to build reactors. The first one went critical in December 1946. They also created a vast mining industry to mine uranium. The successful test of what we called 'Joe-1' occurred at 7 am on 29 August 1949. It had about the same explosive yield as the Nagasaki bomb on which it had been modeled.

President Truman's first reaction was that it was a fake—that those 'Asiatics' could never build an atomic bomb. But once the radioactive fallout drifted to where it could be observed, it was clear even to Truman that those 'Asiatics' had detonated a bomb. On 23 September he made an announcement.

> We have evidence that within recent weeks an atomic explosion occurred in the USSR.
>
> Ever since atomic energy was first released to man, the eventual development of this new force by other nations was to be expected. This probability has always been taken into account by us.

> The recent development emphasizes once again, if such emphasis was needed, the necessity for that truly effective, and enforceable international control of atomic energy which this government and a large majority of the United Nations support.

Crash program

Whatever good intentions Truman might have had in this declaration, they were rapidly overwhelmed by the idea that we were in an existential struggle with the Soviets and that we must stay ahead of them in the nuclear arms race. As I will now discuss, this is what led Truman on 30 January 1950 to issue a public statement that included the words 'I have directed the Atomic Energy Commission to continue its work on all forms of atomic weapons, including the so called hydrogen or super-bomb', a statement widely interpreted as a call for a 'crash program' to build a hydrogen bomb, something which no one at the time knew how to do and about which Truman had no real understanding. It was a terrible decision. If Truman had paid attention to the report of a scientific advisory committee headed by Oppenheimer it might have been avoided.

A hydrogen bomb is so named because its energy source is the fusion of isotopes of hydrogen, a process that can release energy. The most energetically productive reaction is the fusion of a heavy hydrogen nucleus, which consists of one neutron and one proton, with a nucleus of super heavy hydrogen, which consists of one proton and two neutrons, to produce a nucleus of helium and a neutron that carries away most of the energy. To initiate such a reaction one needs an atomic bomb to produce the requisite temperatures. The first suggestion of doing this seems to have come from the physicist Enrico Fermi. He had used his trip to Sweden in December 1938 to collect his Nobel Prize to leave Fascist Italy and come to the United States. Early in 1942 Edward Teller visited him at Columbia University, where he had become a professor, to discuss atomic weapons. Fermi casually suggested that they might be used to create conditions of temperature that pertain in the interior of stars, where fusion reactions provide the energy. Teller at first thought that the idea was crazy, but he soon became obsessed by it. He thought that the fission device was as good as done and that the focus should be on the fusion device. This obsession never left him. During the war Oppenheimer had to create a special division at Los Alamos in which Teller and a small group could work on the 'super'. They found themselves stuck. You could ignite some, but then they would radiate and cool off before the fusion reactions could propagate. It was like trying to light a log with a match. In early September 1945 Fermi gave some summary lectures on the super. They read like a physics text. He argued that he did not see how to make it work. One of the members of the audience was Fuchs.

On 19 September 1945 the British delegation gave a formal lab party. Fuchs was in Albuquerque, where he had gone to meet his Soviet courier Harry Gold. He turned over notes on Fermi's lectures. Again, I do not know in what form. One can find the lectures on the internet, but they seem to be a later version. This was the most significant thing that Fuchs gave the Soviets on the hydrogen bomb. It revealed the calculations on fusing hydrogen isotopes and also the difficulties. Above all it revealed that we had a program; the Soviets began looking into the matter.

In 1946 Truman signed an act that converted atomic energy from a military to a civilian domain. The Atomic Energy Commission was created—now replaced by the Department of Energy. In October 1949 a General Advisory Committee was asked to make recommendations about constructing a hydrogen bomb. Oppenheimer was the chairman and the committee consisted of people like the physicist I I Rabi and Enrico Fermi, as well as the president of Harvard, James Conant. Some years ago, Rabi described for me how it went.

> In fact, the debate was recorded but the tapes were deliberately destroyed soon afterward. It is a great pity that we cannot hear the voices of Fermi and Oppenheimer and the rest in that very fateful discussion [...]
>
> After the Russian explosion [Joe-1, August 1949], certain people thought that we need some counterthrust. And [Ernest] Lawrence, [Luis] Alvarez and Teller felt that the only thing was to go full tilt for the Super. In fact during Los Alamos some people felt that it should have higher priority than the ordinary fission bomb. 'We shouldn't dillydally with the fission bomb but go for the Super'—so said some very eminent physicists. Not very sensible, but very eminent. I wonder what they think about that now. Anyway after the Russian explosion they wanted to go ahead with a specific Super model. This model is what came up before the committee. There were two objections to it. In the first place it was a very chancy thing because, basically, we did not know how to make it. And then, just about this time it was shown that it wouldn't work, it wouldn't propagate—at least not that particular model. But the general kind of thing they were talking about would have been absolutely devastating because it was so big [...]
>
> In any event, there was strong agreement within the committee that we should not go ahead. We all agreed that if it could be made to work it would be a terrible thing. It would be awful for humanity altogether. It might give this country a temporary advantage, but then the others would catch up—and it would just louse up life. Fermi and I said that we should use this as an excuse to call a world conference for the nations to agree, for the time being, not to do further research on this.

Truman decided to bypass these scientists and bring in his own advisors—three non-scientists—and, as Rabi noted, 'it was just about this time that Klaus Fuchs was arrested in London. All of this got to President Truman and built up such a head of steam that he was practically forced to declare that he was going to give the Super top priority [...] I never forgave Truman for buckling under pressure. He simply did not understand what it was about. As a matter of fact, after he stopped being President he still didn't believe that the Soviets had a bomb in 1949. He said so. So for him to have alerted the world that we were going to make a hydrogen bomb at a time when we didn't even know how to make one was one of the worst things he could have done. It shows the dangers in this kind of thing. He didn't have his own scientific people to consult and give him an impartial picture.'

One is led to wonder: what if Fuchs had not given the Soviets what he did? Perhaps they would not have developed the bomb until Truman left office, and might Eisenhower have taken a different route? In 1951 the Polish-born mathematician Stanislaw Ulam and Edward Teller proposed a novel design of a fusion bomb. In essence it was a two-stage device in which, as in the classical super, the first stage was a fission bomb. But the radiation emitted by this explosion was used to compress a container of the fusible material, thus causing it to ignite. Using this design our first successful test of a hydrogen bomb was in 1952. The Soviets followed in 1955, the British in 1957, the French in 1968 and the Chinese in 1967. At first the idea was to make bigger and bigger bombs. And that happened for a while. The biggest test was the Soviet Tsar Bomba in 1962. It was about 3,000 times more powerful than the Nagasaki bomb. But it was realized that there was another use for staged nuclear explosions. One could drastically economize on the plutonium used in fission bombs. The bombs could also be lighter, which is helpful if they are to be put in rockets, and they would be much more powerful than, say, the Nagasaki bomb. They could be built in the thousands, which is what has happened in the United States and the Soviet Union. What strikes me about all of this is that it is pointless. The targets are still too small. There are asymmetric wars all over the world. Can anyone imagine a role for any of these bombs? Human ingenuity has managed to create the most devastating and the most useless weapon ever invented.

Chapter 9

Latent threats

Iran again

The Ayatollah Akbar Hashemi Rafsanjani has had a complex career in Iran. He was born in 1934 to a wealthy family of pistachio farmers. He studied theology with the Ayatollah Khomenei in Quom. When Khomenei took over the government in 1979 Rafsanjani became the first speaker of parliament. During the 1980–88 Iran–Iraq war he was the *de facto* commander of the armed services. Thus he was in a position to know about Iran's nuclear weapons program. Hence he has to be taken seriously when he said recently, 'When we first began, we were at war [with Iraq] and we sought to have that possibility [of nuclear weapons] for the day that the enemy might use a nuclear weapon. That was the thinking. But it never became real.' He then goes on to add, 'We were at war and Iraq had come close to enrichment before Israel destroyed it all. Our basic doctrine was always a peaceful nuclear application, but it never left our mind that if one day we should be threatened and it was imperative, we should be able to go down the other path.'

I will analyze the last two sentences. The first of the two seems to me to be entirely confused. On 7 June 1981 eight Israeli planes destroyed the so-called Osirak reactor near Baghdad. It had been sold to the Iraqis by the French and they named it: Os for the Osiris class and Irak for its destination. It was a small reactor compared, say, to a power reactor. It generated 40 megawatts of power, whereas a power reactor generates a couple of billion. It was fueled by highly enriched uranium—93%—of the kind you would use in a weapon. At the time of the raid the French had supplied 12.5 kg. Not much less than you would need to make a bomb—about 16 kg. (The amount actually depends on the type of bomb, but this gives the order of magnitude.) The Israelis did not seem to pay any attention to this. They were concerned about the reactor's capacity to produce plutonium like their own Dimona reactor. But the Osirak reactor was about as poorly designed for making plutonium as one could imagine. To make plutonium you want the least enriched uranium possible. It is a reaction in which the common isotope U-238 absorbs a neutron that leads to plutonium. This reactor could produce about 0.07 kilograms a year. Hence it would take centuries to produce enough to make a bomb, which requires about ten kilograms. Thus from the point of view of proliferation the Israeli raid was less than useless—a failure, apparently, of the vaunted Israeli intelligence. It left the Iraqis with a substantial amount of weapons-grade uranium and inspired them to begin a uranium enrichment program. A Q Khan tried to sell his centrifuge package to Saddam Hussein, but Hussein did not trust him. Rafsanjani discusses an unsuccessful trip to Pakistan to see A Q Khan to acquire nuclear technology. He does not give the date of this trip, but starting in 1987 and lasting for nearly a decade Khan sold this material to Iran. The Iraqis never got far with their enrichment program.

The sentence 'Our basic doctrine was always a peaceful nuclear application, but it never left our mind that if one day we should be threatened and it was imperative, we should be able to go down the other path' bears a remarkable resemblance to the sorts of things that the Israelis say. But what does it mean? How do you retain the ability 'to go down the other path'? The only reasonable meaning I can attach to this is that you retain facilities devoted to work on nuclear weapons. I very much believe that this is the case and that the Iranians have plans for nuclear weapons—perhaps plans they bought from A Q Khan. It does not take many theoretical physicists to carry out this work. The Iranians have now begun to dismantle their centrifuges. This is a step in the right direction. It will be difficult—perhaps impossible—to learn how far along they are in the conceptual matters. In any event, knowledge is a genie which cannot be put back into the bottle.

One of the remarkable coincidences in the negotiations that led to the Joint Comprehensive Plan of Action—the 2015 agreement on the Iranian nuclear program—was the role of MIT. When in the 1970s the Shah of Iran decided to take Iran nuclear he turned to MIT. He gave an endowment of 20 million dollars to the Institute and funded people from Iran to study there. One of them was Ali Akbar Salehi, who later became the head of the Atomic Energy Organization of Iran. As it happened, Ernest Moniz, the US Secretary of Energy until at least January 2017, was a young assistant professor at MIT at the time. Salehi had come to study nuclear engineering. There is no indication that the two men crossed paths at that time, but one might suppose that this coincidence created some sort of bond. When the Shah fell in 1979 the various Iranian MIT students took different directions—some remaining in the United States and some returning to Iran. The Ayatollah Khomeini, who took over Iran, declared the Shah's program 'haram' (forbidden) since it involved nuclear weapons, and he closed most of it. One of the few survivors was the Teheran Research Reactor that had been purchased as part of President Eisenhower's Atoms for Peace Program. Probably no one bothered to inform Khomeini that it was fueled by weapons-grade uranium. He died in 1989 and was succeeded by Khamenei, and the nuclear program was re-invigorated.

Starting in 1990 the Iranians and the Pakistani proliferator A Q Khan began doing business. Iran bought the prototype Pakistani centrifuges and, reverse engineering them, began their enrichment program. It is an unanswered question as to whether they also bought the plans for a Chinese nuclear device that Khan had sold to the Libyans. Indeed, the precise nature of the military dimension of the Iranian program is still not known despite years of effort by the International Atomic Energy Agency, and despite what is called by the IAEA 'Final Assessment on Past and Present Outstanding Issues regarding Iran's Nuclear Programme' [1], which was issued on 2 December 2015. As far as I am concerned, the military dimension of Iran's nuclear program is still not known. I will explain why and then discuss the question of whether this matters.

A typical paragraph of the IAEA report reads, 'Information available to the Agency prior to November 2011 indicated that in the early 1990s Iran may have received design information for a nuclear explosive from a clandestine nuclear supply network. Iran provided the Agency with a copy of a one-page handwritten document said to be an offer from this nuclear supply network relating to centrifuge technology. During discussion with Iran in 2005 the Agency identified a 15 page document relating to the conversion of uranium compounds into uranium metal ("uranium metal document") and the production of hemispherical enriched uranium metal compounds.'

What is one to make of this? One would imagine that the 'clandestine nuclear supply network' was that of A Q Khan. Either the Iranians bought the weapons design from Khan or they did not, and what is this 'offer'? The Iranians did buy the whole centrifuge package. And what is this '15 page document'? Hemispherical uranium metal is what one would use in the design of an implosion bomb. The Iranians have mastered the techniques required to make metallic uranium, which they have used in fabricating the fuel elements for the Teheran Research Reactor. What does this 15 page document add and what does 'identified' mean? Did the Agency study this document or were its inspectors just shown the cover?

Then there is the matter of the mysterious 'foreign expert' who has haunted these reports like the ghost of Hamlet's father from the beginning. This expert 'worked for much of his career in the nuclear weapon programme in his country of origin'. The Iranians agree that he was employed by them from 1996 to 2001, but say that he was working on 'nanodiamonds'. What else was he working on and who is he? It is unlikely that we will ever find out for sure.

In summary: all the external evidence points to the fact that the Iranians have attempted to design a uranium nuclear implosion device. It seems clear that the chemical explosive implosion techniques were being studied at the military site Parchin near Teheran. The International Atomic Energy Agency was never allowed a real inspection of this site, but something about it was learned from satellites. In preparing the report cited above, the Agency was allowed a brief inspection and of course found nothing. The satellites had shown a significant clean-up effort. The Iranians have attempted for many years to buy things like explosive bridge wire detonators under the guise they were needed for mining, when it is clear to anyone with any knowledge of the subject that their only use is in implosion devices. The new rationale that the Iranians provided for this report was the prevention of explosive accidents and they conveniently cited five examples which are not given in the report.

In reading this IAEA report, I at least have the impression that the Agency has thrown up its hands. It has now closed its books. It was given an impossible task and it did what it could. The real question is: does it matter? I think not, provided the rest of the Joint Agreement is taken very seriously. Whatever clandestine knowledge of the design of nuclear weapons the Iranians acquired will not help in their construction if they are denied the fissile material. That matters.

A Saudi bomb

Prince Turki bin Faisal Al Saud, the former Saudi intelligence chief, announced in 2015 at a conference that dealt with nuclear matters and was held in Seoul, South Korea, 'Whatever the Iranians have, we will have too.' He is an interesting man. He was born in 1945 in Mecca. Until he was 14 he was educated in Saudi Arabia, but then he was sent to Lawrenceville School in New Jersey, which is a kind of feeder school for Princeton. But instead of Princeton he went to the Edmund A Walsh School of Foreign Service at Georgetown University, where he was a classmate of Bill Clinton. In interviews he has said that his Lawrenceville classmates had difficulty with his first name, so they called him 'turkey'. After returning to Saudi Arabia he held the position of head of Saudi intelligence for 23 years. In 2005–6 he was ambassador to the United States. He had a falling out with some members of the royal family and in recent years has been the chairman of the King Faisal Center for Research and Islamic Studies. It is not clear what this center does and there is no indication that Turki has any scientific training.

Compared with Iran, physics in Saudi Arabia is in its infancy and much of its nuclear program is aspirational. They hope to construct 16 nuclear power reactors by 2040, with the first one going on line in 2022. In 2007 the CERN laboratory made its first long-term visitor appointments from Saudi Arabia. One of them was a woman named Nadar S Alharbi. In the spring of 2014 the American physicist Sultana N Nashar, herself a Muslim, was able to visit four Saudi universities. There are now some 32 public universities and ten private universities in the kingdom. Most of them are co-ed, except for Princess Nora University for women and a few for men, which are converting. In the women's colleges western dress is worn and hairstyles vary. But once outside all the women wear the black abaya. There seem to be several hundred physics students in these universities, but I have not been able to look at the curricula. From what I can see Saudi Arabia simply does not have an impressive scientific infrastructure.

Let us try to see what Turki's statement means. If the present indications hold up Iran will be left with some 5,000 gas centrifuges entirely of their own manufacture. As far as I can tell Saudi Arabia has none. Iran has several functioning nuclear reactors. Saudi Arabia has none. Iran has a facility at Isfahan for converting enriched uranium hexaflouride into metal. Saudi Arabia has none. The only thing that Saudi Arabia has in abundance is oil money. The only way it can carry out Turki's proposal to match the Iranians is to buy it.

Ukraine

Watching the 2014 crisis in Ukraine unfold, it was easy to forget how much worse it could have been. In the sense of civil power, Ukraine is a nuclear state. Recall that the Chernobyl disaster took place at Pripyat, in northern Ukraine, some 60 miles from Kiev. Ukraine still has 15 operating nuclear reactors, which contribute about half the country's electricity. The enriched uranium used to fuel these power reactors is largely provided by the Russians (some is provided by General Electric). The Russians presumably have also been removing the plutonium the reactors produce. But it is Ukraine's history as a military nuclear power that is so striking. In 1991 Ukraine had the third largest nuclear arsenal in the world—only the United States and Russia had more.

At the end of the Cold War, Ukraine had some 1,900 strategic nuclear warheads and hundreds of tactical nuclear weapons—the kind that are designed to be used on the battlefield. Many of these were stored in the Crimea: one of the Soviets' most important weapons sites was at Krasnokamenka, in the Kiziltashsky valley region of Crimea, where a secret underground facility was used to assemble and store nuclear warheads, some of which were then sent to other Warsaw Pact countries. As part of the Soviets' military defenses against Europe the tactical nuclear weapons were distributed to Belarus, Kazakhstan and Ukraine, as well as Russia. After Russia, Ukraine, which was seen as a primary front in any battle with the West, got the most.

It has never been clear how to use so-called tactical nuclear weapons, which were promoted for use on advancing troops and tanks. As Oppenheimer said of the hydrogen bomb, the military targets are too small, and in practice, such weapons would produce collateral damage to your own troops. But it is clear how to use the strategic weapons, which are designed to attack whole cities: you put them on missiles or bombers. The Ukrainians had dozens of both.

Of course, at the time this arsenal was built, during the Cold War, Ukraine was part of the Soviet Union, so these weapons, along with Russia's, were part of a single stockpile. The newly independent country of Ukraine inherited them. In 1990 the Ukrainian parliament adopted a non-nuclear policy, which included ridding the country of these devices. In 1992 then-prime minister and later president Leonid Kuchma decided that nuclear disarmament was not such a good idea after all and that Ukraine should keep a stockpile of nuclear missiles. He did not get his way, however, because in 1994 the parliament voted to make Ukraine a nuclear weapons-free state, and that year the Ukrainians signed the so-called Budapest Memorandum, in which they agreed to dismantle their nuclear stockpile.

In view of Russia's recent annexation of Crimea, it is worth noting that the Budapest accord, which was signed by Russia itself along with Ukraine, the US and the UK, required the signers to make security assurances that any incursion on Ukrainian territory by a nuclear power would be promptly taken before the United Nations Security Council. Indeed, Ukraine's newly appointed Prime Minister, Arseniy Yatsenyuk, has cited this agreement in relation to Russia's intervention, calling on the Kremlin not to 'violate the Budapest Memorandum'.

The memorandum, however, did not make security guarantees to this effect, which would have forced the UN to take action and put Russia in a very tight spot.

In the event, following the 1994 agreement, trains began moving Ukraine's nuclear stockpile to disarmament facilities in Russia. About 5,000 nuclear-related devices were moved out on some 100 trains. The operation was completed in 1996 and Ukraine joined the small club of nuclear states, which now includes Libya and South

Africa, which have voluntarily given up their nuclear arsenals. In the 1990s, Belarus and Kazakhstan also gave up their weapons. But the vast store of tactical and strategic bombs turned over by Ukraine was by far the largest in this group. In 2012 the last of Ukraine's supply of highly enriched uranium was turned over to Russia.

As for the civilian nuclear power program, there are plans to build 11 new units, and some of the older ones are being decommissioned. There is always a risk with nuclear reactors, but hopefully the lessons of Chernobyl have been absorbed. The reliance on Russian uranium for these reactors is another way the Kremlin could continue to exert leverage over Ukraine, although some of the fuel is being supplied by the United States and one imagines this could be increased. Without these reactors Ukraine would run the risk of going dark. (Since the Russian incursion in the Crimea, Ukraine's interim president has said the country would deploy armed forces around its nuclear facilities, while the Ukrainian parliament has called for international monitors to help guard its reactors.)

We can debate what might have happened if Ukraine had kept at least part of its nuclear arsenal, as Kuchma wanted. Would this have deterred the Russians or would we be facing a nuclear war? Or might the weapons themselves be in danger of falling into the wrong hands during an insurrection? Fortunately for all of us these questions are now academic.

Terrorists

From what I have said so far it is clear that the idea of a terrorist group like Al Qaeda or ISIS constructing a nuclear weapon from scratch is absurd. On the other hand, they might buy one. Pakistan is often mentioned as a seller, or perhaps North Korea. It is the case that the North Koreans were engaged in helping the Syrians build a reactor. The Israelis took care of that one in 'Operation Orchard' in 2007. From what I have written so far it is clear that Pakistan is a worry and close attention must be paid. It is not something I put a high priority on. How would a group like ISIS deliver such a weapon? I do not lose sleep over a high explosive bomb that might spread radioactive isotopes, a so-called 'dirty bomb'. Such a thing might cause some short-term panic, but the effects would not be long lasting or widespread. I am always reminded of something my late friend and colleague Robert Serber told me about. He was at Los Alamos during the war and soon after visited Hiroshima. He brought a Geiger counter to measure radioactivity. There was only one place in the city where he found any. This was the ruins of a hospital and the light radioactivity was coming from isotopes that had been used in medical procedures. On the other hand, if a group like ISIS could purchase a bomb they could deliver it in a truck and destroy a city like Baghdad. This is still another reason to eradicate this group.

Reference

[1] IAEA 2015 *Final Assessment on Past and Present Outstanding Issues regarding Iran's Nuclear Programme* GOV/2015/68 (Vienna: International Atomic Energy Agency)

Chapter 10

Oppenheimer's lives

That is what novels are about. There is a dramatic moment and the history of the man, what made him act, what he did, and what sort of a person he was. That is what you are really doing here. You are writing a man's life.

I I Rabi

On Monday 12 April 1954, J Robert Oppenheimer appeared in room 2022 in a two-story building near the Washington Monument, where the Atomic Energy Commission had its quarters. The room had been transformed into a courtroom, and Oppenheimer was subjected to a hearing in which his fitness to have access to classified information—his security clearance—was to be decided. To many of us who had had no previous notice, this was a shock. Apart from anything else, Oppenheimer had been at Los Alamos and was responsible for *creating* much of this classified information, which led to creation of the atomic bomb. He was a towering figure who had been entrusted with the greatest secret scientific project in the country's history. He had been hailed as a hero, with his picture on the cover of magazines like *Time*. He was also the teacher of many of our teachers.

It was some time before the full transcript of the hearing was released. It reads like a great play and in fact has been made into plays. In the end, Oppenheimer lost his clearance. People who had known him from before said that he also lost a good deal of his spirit. The cast of characters at his hearing included some of the most important physicists of the time, such as Edward Teller, who testified against Oppenheimer, and I I Rabi, who thought the proceedings were a disgrace and said so.

An unlikely individual who figured in the hearing was a professor at Berkeley, where Oppenheimer taught, named Haakon Chevalier. The two met in 1937 when Oppenheimer was 33 and already acknowledged to be one of the best theoretical physicists of his generation. He held joint appointments at Berkeley and Caltech and had a comfortable private income from his family. Chevalier, who had been born in New Jersey of mixed French and Norwegian parents, was teaching French literature at

doi:10.1088/978-0-7503-1308-7ch10

Figure 10.1. Young Robert Oppenheimer. This image has been obtained by the author(s) from the Wikimedia website https://en.wikipedia.org/wiki/J._Robert_Oppenheimer#/media/File:Robert_Oppenheimer_1946.jpg, where it is stated to have been released into the public domain. It is included within this article on that basis.

Berkeley. He had written an important book on Anatole France and had translated André Malraux. Later he would write novels. His description of the Oppenheimer of 1937 is written with a novelist's eye. In it I recognize the Oppenheimer whom I got to know 20 years later.

Chevalier writes, 'I can remember my earliest impressions of him at this stage. He was tall, nervous and intent, and he moved with an odd gait, a kind of jog, with a great deal of swinging with his limbs, his head always a little to one side, one shoulder higher than the other. But it was the head that was most striking: the halo of wispy black curly hair, the fine sharp nose, and especially the eyes, surprisingly blue, having a strange depth and intensity, and yet expressive of candor, that was altogether disarming. He looked like a young Einstein, and at the same time like an overgrown choir-boy. There was something both subtly wise and terribly innocent about his face. It was an extraordinarily sensitive face, which seemed capable of registering and conveying every shade of emotion. I associated it with the faces of apostles, either imagined or remembered from Renaissance paintings. A kind of light shone from it, which illuminated the scene around him.'

Of course, 20 years later his hair had grown white and was cut very short. He sometimes had a look of martyrdom, and he could be quite cruel when he thought his time was being wasted. He was no longer really doing physics, but what struck me was how important other physicists—many who had been much more creative than he—felt that his approval was.

What brought Oppenheimer and Chevalier together initially was their common interest in left-wing causes. In reading Chevalier himself as well as what has been written about him, it is never clear as to whether he was or was not a member of the Communist party; but from what I can deduce, the answer was almost certainly yes. In any event, he was a fellow traveler, *pure et dure.* Oppenheimer, on the other hand, until sometime shortly before he met Chevalier, had had an almost total indifference to politics. He reports that until 1936 he had never voted in a presidential election.

He was not aware of the Great Depression until some of his gifted students were not able to get jobs. Even though he had earned his PhD in Germany in the late 1920s, he was more or less indifferent to Hitler until his own relatives had to flee Germany in the mid-1930s. He reports that he never read a newspaper or a news magazine and did not even have a radio or telephone. Yet from 1936 until Pearl Harbor he suddenly had an intense period of political activity, largely in the service of causes, such as that of the Loyalists in the Spanish Civil War and the organization of migratory workers that was closely linked to Communist interests. What brought about the change? At his hearing before the Personnel Security Board of the Atomic Energy Commission in 1954, Oppenheimer suggested an answer.

He recalled that in the spring of 1936, some friends had introduced him to a woman named Jean Tatlock, the daughter of an English professor at Berkeley. She was studying psychiatry. They began a stormy affair, which lasted until he married Kathryn ('Kitty') Puening in November 1940. Tatlock was a member of the Communist party and urgently committed to the activities that Oppenheimer soon became involved with. Through her he met others who were either Communists or fellow travelers. These included Kitty, whose husband, Joe Dallet, had died fighting for the Loyalists in Spain as part of the Abraham Lincoln Brigade. She had been a member of the party during her marriage to Dallet. About this time, and against Oppenheimer's advice, his brother Frank and his wife joined the party, as had several of Oppenheimer's students. This raises the obvious question of whether Oppenheimer himself, despite unambiguous and consistent denials, had been himself a member.

From Chevalier's account, his friendship with Oppenheimer and later with Oppenheimer's wife became very close. Oppenheimer had a deep interest in literature. He once told me that he had read Proust's entire novel by flashlight while taking a long bicycle tour around Corsica, an anecdote that Chevalier relates in his book. During their affair, Tatlock introduced him to the love poetry of Donne. The historian Greg Herken claims that it was with Tatlock in mind that Oppenheimer called the first atomic bomb test, which took place in New Mexico on 16 July 1946, 'Trinity'—this from Donne's fourteenth 'Holy Sonnet': 'Batter my heart, three personed God …'[1]. Herken thinks that naming the test Trinity was a 'secret tribute to Jean Tatlock who had committed suicide in her apartment the previous January'. He is off by a year. After taking sleeping tablets, Tatlock drowned herself in the bathtub in January 1944. That aside, the evidence Herken cites for this notion does not stand up. When in 1962 General Leslie Groves, who had headed the atomic bomb program, asked him this question Oppenheimer said that he was thinking of Donne's poetry. Oppenheimer wrote, 'I did suggest it […] why I chose the name is not clear, but I know what thoughts were in my mind. There is a poem of John Donne, written just before his death which I know and love. From it a quotation

As West and East
In all flat Maps—and I am one—are one,
So death doth touch the Resurrection.

[1] J Donne 1633 Holy sonnet XIV *Love Songs and Sonnets.*

'That still does not make Trinity; but in another better known devotional poem Donne opens, "Batter my heart, three person'd God". Beyond this, I have no clues whatever.'

The poet Edith Jenkins, who was a friend of Tatlock's, wrote a memoir entitled Against a Field Sinister [1] in which she discusses this. This is what Jenkins writes: 'Jean introduced Oppenheimer to the poetry of John Donne. When the trial bomb went off in the Jornado del Muerto in New Mexico, Oppie dubbed it "Trinity" after Donne's sonnet "Batter my heart, three personed God" [...] That was approximately six months after Jean's death.' I was able to contact Jenkins by e-mail and pointed out the misprint about the date when her friend died. She confirmed to me that she knew of no evidence that the naming of Trinity had anything to do with Tatlock.

In Berkeley, Chevalier tried to introduce Oppenheimer to Malraux, who was visiting California. It was only subsequent to Oppenheimer's hearing that Chevalier was able to arrange a meeting, after which Oppenheimer made one of his typically Delphic remarks: 'Malraux has some understanding as to what science isn't. But he has no conception of what science is.' Malraux later commented that what Oppenheimer should have done at his hearing was to stand up and say *Je suis la bombe atomique!* and then, presumably, walk out.

In late 1942 an event occurred that ultimately ruined both Chevalier's and Oppenheimer's lives. Herken supplies a clear chronology, though exactly who said what to whom is still murky. A British-born petroleum engineer named George Eltenton, who had lived in the Soviet Union and was a member of the party, was approached by the third secretary of the Soviet consul in San Francisco—a man named Pyotr Ivanov—to try to learn which of the physicists working at Berkeley on matters related to nuclear energy might be willing to share information with the Soviet Union. Ivanov knew that Eltenton had a relationship with Chevalier and that he, in turn, knew Oppenheimer. At this point motive and intent become unclear. In his book Chevalier gives his version. He says he received a phone call from Eltenton, who wanted to see him on an important matter. When he went to Eltenton's house he learned that this was to sound him about out about recruiting Oppenheimer. Chevalier says that Eltenton seemed nervous and unsure of himself and that he, Chevalier, told him in no uncertain terms that Oppenheimer would never agree to such a thing. In later depositions Eltenton, for whatever it is worth, confirmed this version of the story. The matter might have ended there except for what Chevalier did next. Soon afterward he and his wife were invited for dinner at the Oppenheimers'. In a private moment Chevalier told Oppenheimer about Eltenton with the intent, he says, of simply warning him about possible espionage. He notes that, as far as he was concerned, this ended the matter. He writes, 'We went back into the living room with the cocktail shaker, the gin and the vermouth, and joined our wives. I dismissed the whole thing from my mind.'

Meanwhile an officer in the Army Counter Intelligence Corps named Boris Pash had become involved with the investigation of security for the army. There was a turf war between the army and the FBI, which was engaged in wiretapping everyone who was suspected of spying for the Soviets, including, eventually, Oppenheimer. By the spring of 1943, Oppenheimer had moved to Los Alamos as its director, but in

August he came back to Berkeley for a visit. He casually mentioned to an army security officer that he had been contacted some months earlier about possible espionage. Promptly he found himself confronted by a formal inquiry conducted by Pash, who recorded the colloquy.

It was at this point that Oppenheimer made a dreadful mistake. Instead of telling Pash the simple truth, assuming that it was the simple truth, he invented a story. Later he said he was trying to protect Chevalier while at the same time conveying to the authorities his concern about espionage. He said that he had been approached by 'intermediaries'—plural, whom he would not name—to divulge information about the nuclear program, and he knew for a fact that three other people had been approached. This appears to have been a total fabrication. Upon further questioning, Oppenheimer mentioned Eltenton as one of the intermediaries. He hinted that a member of the Berkeley faculty had been involved but refused to give the name.

Pash reported the matter to General Groves, and he and his security officer, John Lansdale, interrogated Oppenheimer soon afterward during a train trip. Oppenheimer still did not reveal the name of Chevalier, and Groves did not order him to do so. This changed the following December when Groves did insist on having the name of Oppenheimer's contact. He gave Chevalier's name, adding that he was 'quite a Red'. But instead of relating to Groves what was probably the true version of the story—namely Chevalier's—Oppenheimer invented yet another fabrication. This time he admitted that there were not three people involved but, he said, only his brother Frank.

As Oppenheimer said in the hearing:

> The story I told Pash was not a true story. There were not three or more people involved. There was one person involved. That was me. I was at Los Alamos. There was no one else at Los Alamos involved. There was no one at Berkeley involved. When I heard the microfilm or what the hell, it didn't sound to me as to this reporting anything that Chevalier had said, or at that time the unknown professor had said. I am certain that was not mentioned. I testified that the Soviet consulate had not been mentioned by Chevalier. That is the very best of my recollection. It is conceivable that I knew of Eltenon's connection with the consulate, but I believe I can do no more than say the story told in circumstantial detail, and which was elicited from me in greater and greater detail during this was a false story. It is not easy to say that.
>
> Now, when you ask for a more persuasive argument as to why I did this than that I was an idiot, I am going to have more trouble being understandable.

Now Oppenheimer had put two people in jeopardy, Chevalier and his own brother. Chevalier was pursued intensively, finally losing the opportunity to work in this country or even hold an American passport. He took advantage of his dual citizenship and moved to France. Frank Oppenheimer was equally harassed out of academic life and, for a while, took up ranching.

This in essence is the 'Chevalier incident'. But the questions are, why did Oppenheimer behave in this way, and why was he unable to defend himself more effectively at his hearing? On the latter I may be able to shed some light. In 1958–9 when I was at the Institute for Advanced Study in Princeton, of which Oppenheimer was the director, I once shared a train ride with him to New York. He suddenly began discussing his 'case'. He told me that he found it so unbelievable at the time that he thought it was happening to someone else. On the former question, I think he was caught in the dilemma of having to choose between protecting a friend and protecting his country. He thought, mistakenly, that he was clever enough to do both.

Bearing on Oppenheimer's plight was that he had also made enemies. One of them was a Berkeley colleague, the physicist Ernest Lawrence, inventor of the cyclotron, who was persuaded that Oppenheimer was immoral. This came about because of Oppenheimer's relationship with the wife of another colleague, Richard Tolman. Lawrence learned at a cocktail party that Oppenheimer had had an affair with Tolman's wife, Ruth. Lawrence then passed this information to Lewis Strauss, a member of the Atomic Energy Commission and President Eisenhower's adviser on atomic energy. Strauss added it to his catalogue of reasons for hating Oppenheimer, who had made a fool of him in a congressional hearing. Oppenheimer had testified in favor of offering small amounts of radioactive isotopes to friendly countries for research, as a goodwill gesture. Strauss had opposed the idea, though his fellow commissioners supported it. When asked if the isotopes could be used for military applications of atomic energy, Oppenheimer replied, 'No one can force me to say that you cannot use these isotopes for atomic energy. You can use a shovel for atomic energy; in fact you do. But to get some perspective, the fact is materials have played no significant part, and in my knowledge, no part at all.' It was not only what he said but the way he said it.

In any event, Strauss noted the piece of intelligence from Lawrence in a memorandum to himself. He remarked that Oppenheimer 'first earned [Lawrence's] disapproval a number of years ago when he seduced the wife of Prof Tolman at Caltech. According to Dr Lawrence it was a notorious affair which lasted for enough time for it to become apparent to Dr Tolman, who died of a broken heart.'

Clearly there are two questions to be distinguished here. The first is, did Lawrence believe this? And the second is, is the story true? I have no reason to doubt that Lawrence believed it, and I have no doubt that this added to the reasons why he, Strauss and Edward Teller, who was also informed, wished to destroy Oppenheimer. But I have several reasons for believing that it was not true, at least in the way that Lawrence described it. I decided to try to confirm or deny it. This is not easy. All the principals are dead. Tolman died in 1948 and his wife Ruth in 1957, while Lawrence died in 1958 and Oppenheimer in 1967.

Not many witnesses were left from the prewar years when this affair supposedly took place—but there were a few. Among them was the physicist Robert Christy, who became Oppenheimer's student in 1937. He died in 2012. When I asked Christy, he said that he had never heard of an affair between Oppenheimer and Ruth Tolman, though it was well known on the campus that Robert and Kitty had had an affair while Kitty was still married to Richard Stewart Harrison, a British doctor who had become her husband in 1938, after Joe Dallet died. This was confirmed by Tom Tombrello, the chairman of the physics, mathematics and astronomy divisions at Caltech. Tombrello came there in 1961, when the premarital relationship between Oppenheimer and Kitty was still being discussed. He told me that he had actually worked with Harrison, who later remarried and then led an entirely happy life. Moreover, Tombrello was able to discuss this with Marge Lauritsen Leighton, the widow of Tom Lauritsen, a Caltech nuclear physicist. She said that when she was first married to Lauritsen they lived in the Tolmans' house, with the Tolmans, and that she was a close friend of both the Tolmans and the Oppenheimers. She had never heard of any affair between Ruth Tolman and Oppenheimer, and doubted if it had ever happened. Christy also told me that when Oppenheimer used to come to Caltech to teach before the war, he always stayed in the Tolmans' house on campus, and that this continued even after he and Kitty were married. The least one can say is that this affair, if it ever existed, was scarcely 'notorious'. Everyone agreed that no evidence suggests that Tolman died of a 'broken heart'. It was rather a stroke that killed him aged 67. Remarkably, his attending doctor was Kitty's ex-husband Stewart Harrison, who was called in as a friend of the family.

On the other hand, although it seems that Ruth Tolman destroyed many of her personal papers, a little of her correspondence with Oppenheimer is preserved. Two letters—undated—contain passages which on their face indicate an intimacy between them. It could be interpreted as something beyond a close friendship, but who knows? In one of them, Tolman writes, 'The precious times with you that week and the week before keep going through my mind, over and over, making me thankful but wistful, wishing for more. I was grateful for them, Dear, and as you know, hungry for them too.' Later in the letter there is an affectionate reference to her husband. In another she writes about a possible arrangement by which they might be able to see each other. She describes driving to San Diego and seeing 'the long stretch of beach where the sandpipers and gulls played. Oh, Robert, Robert. Soon I shall see you. You and I both know how it will be.' She describes the guest list for a party she is planning to give then. Christy is mentioned as one of the guests, but he was not sure that he could remember the occasion.

How one puts all of this together I am not sure. But, to repeat, the affair, if it happened, was scarcely 'notorious', and it did not seem to break anyone's heart. Incidentally, Edith Jenkins provides a remarkable insight. She writes, 'Oppie was already married to Kitty. I asked Jean [Tatlock] if she regretted refusing to marry him, and she said yes, she would not have done so had she not been so mixed up. I recall responding perhaps that she perceived him as essentially nonsexual. Jean put her cheek against Margy's [Jenkins's child] and held gently the baby hand that was pulling her hair and said, "Maybe you're right…"'.

Was Oppenheimer in fact a member of the Communist party? If this is true, it would mean that he had lied repeatedly to various security agencies in an entirely unambiguous way—an incredibly serious charge. On what is the accusation of party membership based? Once again we have Chevalier to thank. During his prewar Berkeley association with Chevalier, Oppenheimer attended meetings of a small group of left-leaning people, mostly if not entirely Communists. He called this a 'coffee klatsch'. In July 1964 he received a letter from Chevalier with the salutation 'Dear Opje'. ('Opje', a Dutch diminutive, stood for 'little Op'. This is how he was known before the war. 'Oppie' came later.) Chevalier begins by saying, 'I had not thought ever to write you again. Our ways have parted so utterly.' But he had changed his mind. He was writing a memoir—as opposed to a novel—in which he intended to discuss his early years with Oppenheimer. Then comes the bombshell: 'The reason for my writing you is that an important part of the story concerns your and my membership in the same unit of the CP from 1938 to 1942.' In other words, he is saying that Oppenheimer was a member of the party in the very years when he was already engaged in working on nuclear weapons. He seems to think that Oppenheimer has nothing 'to be ashamed of' and that it would be all right to publish this unless Oppenheimer had 'any objections'.

In writing this letter it is difficult to tell whether Chevalier is being naive or malicious. Did he not understand that for decades Oppenheimer had given sworn statements that he had never been a party member? To have been caught lying about this would not only have destroyed Oppenheimer's reputation, but it would also have subjected him to serious criminal charges.

A few weeks after receiving Chevalier's letter, Oppenheimer replied. He denied in the strongest terms that he had ever been a member of the party. The matter was turned over to his lawyers, and Chevalier's accusation was never made in print. In short, we are at an absolute impasse. There is no reason whatsoever, in my view, to believe Chevalier's assertion as opposed to Oppenheimer's denial. Unless someone can come up with real evidence, I take Oppenheimer's word.

What then accounts for the twists and turns of Oppenheimer's security problems? He was a puzzle even to people who knew him well. Nothing was ever simple. Take his name. When asked about his name J Robert, Oppenheimer always said that the J stood for nothing. The fact is that in official FBI documents, and documents involving General Groves, Oppenheimer is referred to as 'Julius Robert Oppenheimer'. There is no mystery. One need only go to the Manhattan Bureau of Records and obtain a copy of Oppenheimer's birth certificate. This document, which I have in front of me, says that Oppenheimer was born 22 April 1904 at 200 West 94th Street in Manhattan, the family residence. It gives his father's age as 33 and his father's birthplace as Germany. His mother Ella, née Friedman, was born in the United States. The name on the birth certificate is Julius Robert Oppenheimer. Why does this matter?

I often discussed Oppenheimer with I I Rabi, who had known him for most of his adult life. Rabi's wife had even gone to Ethical Culture School with him. Rabi remarked again and again that Oppenheimer's problem was identity. Once he said to me that Oppenheimer reminded him of a friend who could never decide whether he wanted to join the B'nai B'rith or the Knights of Columbus. Another time he said that Oppenheimer would have been a much greater physicist if he had studied Yiddish rather than Sanskrit. If, like Rabi, he had accepted his Jewishness, his personality might have been more coherent. Instead Oppenheimer was always trying to create identities for himself, and early on he decided that 'J Robert' presented a much better *figura* than 'Julius Robert'. (One should not underestimate the climate of anti-Semitism that prevailed at the time Oppenheimer began calling himself 'J Robert', which he was certainly doing in the mid-1920s. His teacher of physics at Harvard, Percy Bridgman, in a letter of recommendation written then to Ernest Rutherford, felt obliged to say, 'As appears from his name, Oppenheimer is a Jew, but entirely without the usual qualifications of his race'.)

I think that part of what Lawrence called Oppenheimer's 'left-wandering' was just that—a search for identity. For that reason it was easy for Oppenheimer to take on an entirely different identity once he became deeply involved with the bomb. He did not find it difficult to support Groves when he tried, unsuccessfully, to make Los Alamos a military base with everyone in uniform, including Oppenheimer, who would have had the simulated rank of lieutenant-colonel. This persona had no difficulty telling Groves that his good friend Chevalier was 'quite a Red' or that his brother was involved in an espionage attempt. Once the bomb was built he found it easy to become a 'smiling public man' who referred to General George C Marshall, army chief of staff, as 'George'. And, after the hearing, there was a new persona—the martyr.

I saw these different identities in action on one remarkable occasion. It was a Saturday morning, and I was working in my office at the Institute for Advanced Study. Suddenly there was an influx of TV cameras followed by Oppenheimer and the television journalist Howard K Smith, all heading for our colloquium room. I followed. Oppenheimer said I could stay as long as he did not see me—a strange request. I was able to hear what ensued. When the cameras were rolling, he spoke in that subdued and mesmerizing way about his case and his lack of influence on nuclear affairs. But when the cameras stopped, his voice changed and he talked with Smith about the best New York French restaurants and sailing in the Caribbean. Which was the real Oppenheimer? Probably both—or neither.

I think that I got to know Oppenheimer somewhat better than most of my fellow junior cohorts. In the first place, I stayed at the Institute for two years, which was a little unusual. Then, too, Oppenheimer seemed to find me interesting. He was apparently especially interested that I had gotten to know Reinhold Niebuhr and his wife Ursula, who were also at the Institute. Indeed, in a letter of recommendation for me, which I saw years later, he mentioned this. It also led to one of the oddest moments during my stay: a lunch with W H Auden, the Niebuhrs and Oppenheimer.

Oppenheimer was a great admirer of Auden's poetry. I witnessed and indeed was somewhat responsible for the only encounter of the two of them. When I first came to the Institute in the fall of 1957, as a postdoctoral, I noticed than one of the other visitors was Reinhold Niebuhr. As it happened, Niebuhr was a hero of mine whom I had heard preach several times at Harvard. He was also a friend of my father's. Thus when I met Niebuhr in the Institute cafeteria I introduced myself and received from his wife Ursula a cordial invitation to visit them. Later she confided in me that Reinhold had recently suffered a heart attack and that seeing young people cheered him up. One of my visits was just after I had heard Auden, who was visiting Princeton, give a reading. I told the Niebuhrs about this and was surprised to learn that they were very close friends of Auden—he had even dedicated his poetry collection *Nones* to Reinhold and his wife. They did not realize that Auden was in Princeton.

Not long afterward I found myself on the train to New York, as it happened seated next to Auden. I told him about the Niebuhrs and he seemed quite interested. I then forgot about it until some days later when I got a phone call from Oppenheimer's secretary asking me to come to his office for a luncheon appointment. Not knowing what to make of this, I was astounded to find assembled in his office Reinhold and Ursula Niebuhr, the British historian and Institute professor Sir Llewellyn Woodward and his wife, Oppenheimer and Kitty, and, of course, Auden. We then marched up to the Institute cafeteria, where we were seated at a central table and served. When I walked in I noticed various colleagues such as Freeman Dyson giving me rather fishy looks. I recall the seating arrangement. I was next to Niebuhr, and Ursula was next to him, with Auden at her side. Oppenheimer was across the table. I waited for some sort of epiphany, but the only part of the conversation I can remember was a pointless anecdote that Oppenheimer told about having studied Sanskrit with Arthur Rider in Berkeley.

I do not know which of his personae was on display that day, but it seemed to me like a lost opportunity, or nearly. After lunch I took Auden to meet Dyson, and the two of them rapidly became absorbed in playing word games on Dyson's blackboard. I never saw Auden again, and neither did Oppenheimer. However, this lunch seemed to have had an effect on Oppenheimer. A few years ago some colleagues of mine sent me a copy of a letter of recommendation Oppenheimer had written, unknown to me, to Ragnar Rollefson, who was then chairmen of the physics department at the University of Wisconsin. The letter is dated 4 December 1958. The first two paragraphs make a much higher estimate of my scientific abilities than I would have made myself. But the last paragraph is truly remarkable.

> He [Bernstein] is also a very appealing man. A good expositor, very responsible in his work, a hard worker, and a scholar. He has wide cultural interests, and I believe that apart from the physicists, his best friends at the Institute are the Reinhold Niebuhrs. He would be a great asset to your department.

What this has to do with a physics department is beyond me.

Not long after Oppenheimer's death from throat cancer on 18 February 1967, I attended a memorial service in Princeton. It was a beautiful occasion with talks by Hans Bethe, Henry De Wolf Smyth and George Kennan. There was also music by the Juilliard String Quartet and a recording of Stravinsky's *Requiem Canticles*. I was surprised to see on the program that the music had been selected by George Balanchine. As it happened, I saw Balanchine at the Pilates gymnasium the Monday after the service. I told him that I had not realized he was a friend of Oppenheimer's. He said he that was not and was very surprised when he was asked, apparently by Oppenheimer, to select the music.

This was such an odd recollection that a couple of years ago I decided to try to find out if it really happened. After some considerable detective work I determined what had occurred. In the 1950s something called the Congress of Cultural Freedom was formed. Its purpose was to provide a liberal, somewhat left-wing answer to Communist ideology. Oppenheimer joined, as did Nicolas Nabokov, Vladimir's cousin, who eventually became its secretary-general. In 1959 the Congress held a meeting near Basel at which Oppenheimer spoke. Among other things, he told the delegates: 'I find myself profoundly in anguish over the fact that no ethical discourse of any nobility or weight has been addressed to the problem of the new weapons, of the atomic weapons […] What are we to make of a civilization which has always regarded ethics as an essential part of human life, and which has always had in it an articulate, deep, fervent conviction, never perhaps held by the majority, but never absent: a dedication to "ahimsa", the Sanskrit word that means "doing no harm or hurt", which you find in the teachings of Jesus and Socrates—what are we to think of such a civilization, which has not been able to talk about the prospect of killing almost everybody except in prudential and game-theoretic terms?' Oppenheimer and Nicolas Nabokov became friends and in 1967, when Oppenheimer was dying of cancer, Nabokov was living in Princeton. He was working with Balanchine on a ballet, and he must have discussed with Oppenheimer the possibility of having Balanchine select the music for the memorial service, which is what happened. In thinking about this, two things struck me. The first was that even in death Oppenheimer had assumed another persona, that of Balanchine. The second was something that neither Nabokov nor Balanchine knew: the Congress of Cultural Freedom had been funded by the CIA.

Finally, it is worth noting briefly how good a physicist Oppenheimer was. Most people would agree that his best work was done in 1939 when he and a student, Hartland Snyder, created the theory of what later would be called a black hole. Curiously, he never showed any interest in this work. John Wheeler, the Princeton physics professor who coined the name 'black hole' once told me that at a meeting he tried to talk to Oppenheimer about black holes. Oppenheimer simply turned around and walked away. Black holes were then not fashionable and Oppenheimer was above all a man of fashion. For him the black hole work was something any graduate student could have done. After this the war came, and Oppenheimer never did any significant pure physics thereafter. When I knew him at the Institute he was certainly not doing physics, which did not stop him from making acerbic and often irrelevant comments in our seminars. What he certainly did do was to create a great school of theoretical physics in the United States. He worked closely with experimenters, which was something that stood him in good stead at Los Alamos, where he understood everything from the machine shops to abstract theory. What he was never able to do was to make the great theoretical advance which his brilliance might have permitted. When I think of him I often think of the categories of theoretical physicists that were invented by the Soviet physicist Lev Landau. The worst, Landau thought, was to have a small head and a small bottom. You would have no ideas and no patience to work them out. The best was to have a big head and a big bottom. Then there was Oppenheimer, who had a big head but too small a bottom.

Reference

[1] Jenkins E A 1991 *Against a Field Sinister* (San Francisco, CA: City Lights Publishers)

Chapter 11

L'envoi

In the first chapter I described my brief encounter with nuclear weapons in the summer of 1957. My second and last encounter began the following summer. In the spring of 1958 I was finishing the first of my two years at the Institute for Advanced Study. My best friend at the time was a physicist named Michael Cohen. He had been one of Feynman's rare PhD students at Caltech and could do a wonderful imitation of the master. During Cohen's Caltech years he had been a consultant for the Rand Corporation in Santa Monica and he told me that he had saved Rand vast sums of money by introducing an approximation for calculating something that was vital to the design of nuclear weapons. Cohen said that I should become a consultant for the summer and he would put in a word. So it came to pass that in June I found myself in Santa Monica.

The Rand Corporation was then located in an innocuous looking building near the beach. It had its own cafeteria and tennis court. The place resembled a small college campus except that the only items on the curriculum were nuclear weapons. One of the 'courses' was the selection of targets in the Soviet Union in case we had to retaliate. The security in the physics section was even tighter than at Los Alamos. There was a guard who checked you in and out and another who patrolled the offices to see that no classified document had been left on someone's desk. One of the 'professors' was the late Herman Kahn, who was claiming that we could survive the few mega deaths that a nuclear attack would produce. In *Dr Strangelove*[1], Stanley Kubrick, who had read Kahn, referred to his work place as the 'Bland' corporation. Kahn gave me one of his documents to read. I told him that I found it appalling. That ended that. Edward Teller appeared periodically and the group would examine the latest hydrogen bomb tests in the Pacific with some of the same enthusiasm one would exhibit at a sporting event. There was nothing for me to do, so Cohen and I spent a good deal of time in each other's offices gossiping. At one

[1] S Kubrick 1964 *Dr Strangelove or: How I Learned to Stop Worrying and Love the Bomb* (film).

doi:10.1088/978-0-7503-1308-7ch11

point I was called to add up some numbers dealing with an attempt to make a nuclear shield by detonating atomic bombs in the upper atmosphere. A few years later when I was working at the Brookhaven National Laboratory a copy of the report of which apparently I was one of the authors was sent to the lab for me to review. My Q clearance had lapsed, so I never got to see what was in it.

After some weeks of oscillating between life on the beach and hearing about hydrogen bombs I decided that I had had enough. I had heard that Freeman Dyson, whom I had gotten to know at the Institute, had gone to La Jolla—south of Los Angeles—to consult for the General Atomics Corporation. I later learned that in a previous summer he had invented the TRIGA reactor that was so safe it could be used in hospitals to make radioactive isotopes. I witnessed a demonstration at General Atomics, given in the presence of Niels Bohr. The TRR reactor in Teheran was meant to be a TRIGA, but after the Iranian revolution of 1979 a reactor of a different design was used. I had also learned that Dyson was now working on an atomic bomb-powered spaceship called the Orion. (Kubrick later chose this name for one of his spaceships in *2001: A Space Odyssey*[2].) The idea of using bombs to propel a spaceship had been suggested by some people at Los Alamos, including Stan Ulam (who, with Edward Teller, had come up with the idea that made the H bomb work). Theodore (Ted) Taylor, who had been a premier bomb designer, had left Los Alamos to head the Orion program. I wrote to Dyson to say that I had heard he was working on this project and explaining that life at the Rand Corporation left much to be desired. I was very surprised when he called me a few days later to invite me to come and join him. I happily accepted and a few days later I arrived in La Jolla in my new Morris Minor convertible and with my Q clearance, which was needed to work on anything connected with nuclear weapons.

I told Dyson that the whole idea of setting off atomic bombs underneath a spaceship to propel it struck me as mad. Dyson responded by saying that the internal combustion engine that drives automobiles sounds equally mad. It works because the energy from thousands of small chemical explosions drives the machinery without melting away the engine. That is because each ignition lasts a fraction of a second—not enough time to do any damage. Each small bomb would ablate away some material from the bottom of the ship, but if it was designed right, not enough to destroy it. One of the jobs was to find out how much ablation would be caused by an explosion and this involved studying how the radiation from the bombs was absorbed by the material at the bottom of the ship. This depended on its 'opacity'—its ability to absorb radiation. When I got there Dyson had found a beautiful theorem using quantum mechanics that fixed an upper bound to this. One of my first jobs was to test this numerically to see how quickly this bound was reached. We wrote a couple of papers on this, one of which was published many years later in an astronomical journal. For some time our reports were classified as internal General Atomics documents and we had more or less forgotten about it. Cohen's great triumph at Rand was to devise an approximation method for calculating opacities.

[2] S Kubrick 1968 *2001: A Space Odyssey* (film).

I got a certain perverse satisfaction from informing him that his approximation methods violated Dyson's bound.

I consulted on and off at General Atomics for the next year or so. I finally ended up being paid 35 dollars a day, which was to me a small fortune since my annual stipend from the Institute was less than 5,000. To work on Orion, Dyson took a year's leave of absence from the Institute. The program lasted some seven years before the government turned off the financial spigot. By this time Dyson had become disillusioned, not because he did not think that it would work, but because the radioactive fallout from each mission would cause a cancer death or so. This was my last direct involvement with nuclear weapons. But the experience left me with an interest that has lasted throughout my life.

www.ingramcontent.com/pod-product-compliance
Ingram Content Group UK Ltd.
Pitfield, Milton Keynes, MK11 3LW, UK
UKHW051848100926
8556IPUK00005B/516

9 780750 313094